Oxford Socio-Legal Studies

ENVIRONMENT AND ENFORCEMENT

OXFORD SOCIO-LEGAL STUDIES

Oxford Socio-Legal Studies is a series of books published for the Centre for Socio-Legal Studies, Wolfson College, Oxford (a research unit of the Social Science Research Council). The series is concerned generally with the relationship between law and society, and is designed to reflect the increasing interest of lawyers, social scientists and historians in this field.

Already Published (by Macmillan)

J. Maxwell Atkinson and Paul Drew
ORDER IN COURT: The Organization of Verbal Interaction in Judicial Settings
Ross Cranston
REGULATING BUSINESS: Law and Consumer Agencies
Robert Dingwall and Philip Lewis (*editors*)
THE SOCIOLOGY OF THE PROFESSIONS: Lawyers, Doctors and Others
David P. Farrington, Keith Hawkins and Sally M. Lloyd-Bostock (*editors*)
PSYCHOLOGY, LAW AND LEGAL PROCESSES
Sally M. Lloyd-Bostock (*editor*)
PSYCHOLOGY IN LEGAL CONTEXTS: Applications and Limitations
Mavis Maclean and Hazel Genn
METHODOLOGICAL ISSUES IN SOCIAL SURVEYS
Doreen J. McBarnet
CONVICTION: Law, the State and the Construction of Justice
Alan Paterson
THE LAW LORDS

Already Published by Oxford University Press

Genevra Richardson, with Anthony Ogus and Paul Burrows
POLICING POLLUTION: A Study of Regulation and Enforcement
P.W.J. Bartrip and S.B. Burman
THE WOUNDED SOLDIERS OF INDUSTRY Industrial Compensation Policy, 1833-1897
Donald Harris *et al*
COMPENSATION AND SUPPORT FOR ILLNESS AND INJURY

ENVIRONMENT AND ENFORCEMENT

Regulation and the Social Definition of Pollution

KEITH HAWKINS

CLARENDON PRESS · OXFORD
1984

Oxford University Press, Walton Street, Oxford OX2 6DP

London Glasgow New York Toronto
Delhi Bombay Calcutta Madras Karachi
Kuala Lumpur Singapore Hong Kong Tokyo
Nairobi Dar es Salaam Cape Town
Melbourne Auckland

and associated companies
Beirut Berlin Ibadan Mexico City Nicosia

Published in the United States
by Oxford University Press, New York

British Library Cataloguing in Publication Data

Environment and enforcement.—(Oxford socio-legal studies)
1. Water—Pollution—Law and legislation
I. Hawkins, Keith
341.7'625 K3586

ISBN 0-19-827511-0
ISBN 0-19-827514-5 Pbk

Library of Congress Cataloging in Publication Data

Hawkins, Keith
Environment and enforcement.—(Oxford socio-legal studies)
Bibliography: p.—Includes index
1. Water—Pollution—Law and legislation
2. Law enforcement
3. Sociological jurisprudence
I. Title II. Series
K3590.4.H38 1983 344'.046343 83-8230
342 .446343

ISBN 0-19-827511-0
ISBN 0-19-827514-5 Pbk

Typeset by Word Link Limited, Wembley, Middlesex
Printed in Great Britain at the University Press, Oxford

To my Father, and the
Memory of my Mother

Contents

Preface xi

Acknowledgements xv

PART I: REGULATING POLLUTION

Chapter 1 INTRODUCTION 3

I. Enforcing law 3
II. Enforcing regulation 7
III. A research perspective 15
IV. The formal organization of pollution control 17
V. Postscript 22

Chapter 2 SETTING STANDARDS 23

I. Issues 23
II. Procedure 28
III. Enforceability 32
IV. Postscript 35

PART II: POLLUTION CONTROL WORK

Chapter 3 FIELD STAFF 39

I. Officers 39
II. Doing the job 44
III. Getting your wellies wet 49
IV. Reorganization 53
V. Postscript 56

Chapter 4	THE JOB	57
	I. Autonomy	57
	II. Control	61
	III. Covering	64
	IV. Competence	66
	V. Postscript	70
Chapter 5	CREATING CASES	72
	I. Working definitions	72
	II. Discovery	80
	III. Identification	85
	IV. The organization of discovery and detection	90
	V. Postscript	99

PART III: SECURING COMPLIANCE

Chapter 6	COMPLIANCE STRATEGY	105
	I. Compliance	105
	II. Images of polluters	110
	III. The enforcement game	118
	IV. Bargaining	122
	V. Postscript	126
Chapter 7	NEGOTIATING TACTICS	129
	I. The array	129
	II. The projection of authority	133
	III. Enforcement moves	141
	IV. Extending control	147
	V. Bluffing	149
	VI. Postscript	153

Chapter 8 PRACTICAL CRIMINAL LAW 155

I. Taking a stat 155
II. Assessing blame 161
III. Credibility 171
IV. Postscript 173

PART IV: LAW AS LAST RESORT

Chapter 9 USING THE LAW 177

I. The process 177
II. Organizing data 183
III. Images of the law 187
IV. Postscript 190

Chapter 10 LAW AS LAST RESORT 191

I. Prosecution *in extremis* 191
II. Efficiency 198
III. Justice 202
IV. Postscript 207

Notes 209

Appendix: A Note on Research Methods 225

Statutes 233

Cases 233

References 235

Author Index 249

Subject Index 252

Preface

They say that anyone found swimming in the Rhine these days is assumed to be attempting suicide. But while the locals may smile when they say this, the remark is meant seriously. The Rhine has had thousands of tons of toxic wastes—chromium, lead, zinc, arsenic, and mercury—dumped into it and is believed to contain more than two thousand different chemicals. This assault on the river is aggravated by heated effluents and discharges of raw sewage. The fish which survive all this are no longer edible, and often covered in ulcers. Yet the Rhine provides drinking water for twenty million people.

Water pollution is already a matter of critical importance; indeed experts in the use and conservation of natural resources predict that water quality will become one of the major issues of the next decade (*Guardian,* 1980; *Washington Post,* 1980). In developed countries pollution of watercourses is all-pervasive, the consequence of domestic, industrial, and agricultural activity. Centres of population produce large amounts of domestic sewage which must be treated before being discharged to rivers. Many industries also consume vast quantities of water in their manufacturing processes, water which must again be disposed of to watercourses after use. In rural areas, meanwhile, farming contaminates water; many farm effluents are extremely polluting; the liquor which is a by-product of silage manufacture, for example, produces an exceptionally powerful organic pollutant which can kill most living things in a stream over a distance of several hundred yards.

But pollution does not, as is sometimes believed, arise simply from the spread of industry and agriculture. Wherever there are people, there is pollution. In economically less developed countries water pollution is a grave matter, and it is believed that at least ten million children die of water-borne diseases every year (*Observer,* 1980).

In Britain pollution has for years been treated as a matter for legal intervention. Originating with a specific concern for the protection of public health from water-borne disease, efforts to control the pollution of water by law have gradually been placed on a broader footing, reflecting changing expectations of our environment. The mandate of the modern pollution control agency has now been enlarged to the point where its duties incorporate water supply, land drainage, and the protection of fisheries, recreation, and amenity.

From the point of view of the sociologist of law, pollution legislation is a typical example of the use of law to regulate economic life, which is often taken to be one of the characteristic features of industrialized societies. Regulation is a means of coping with technological change: specialized enforcement bureaucracies have been invented to get to grips with the problem of order in complex societies. Prohibitions, enforceable by the criminal sanction, have been grafted onto an existing structure of criminal law in an effort to manage economic relationships and further the protection of the public by the control of certain forms of behaviour regarded—in excess—as harmful or undesirable. Existing forms of legitimate conduct have been transformed into deviant ones in an attempt not to repress the activity, but to regulate it.

Regulation now involves an enlargement of public authority over wide tracts of organizational and business life—though some would argue that its impact is more symbolic than concrete (e.g. Edelman, 1964). This view, however, is not shared by those with a practical interest in water pollution control, judging by the remarks of pollution control staff and industrialists, both of whom hold the view that in general the design and enforcement of regulations is crucially important in shaping the conditions under which manufacturers compete in the domestic and international market-place. The policies adopted and the practices enforced by the pollution control authorities are believed to influence industrial growth, productivity, and profitability.

This book is a sociological study of the regulatory process and explores a small part of the implementation of social and economic regulation. It analyses how legal rules are enforced when those rules seek to regulate the economic and social use of physical resources. It is also an enquiry into the nature of the discretion involved in law enforcement work. Specifically, the book is about the part played by criminal law in the daily routine of protecting water quality, and how, in a bureaucratic system of administrative enforcement, the terse, abstract statements of law are translated into behaviour. Since the routine behaviour of pollution control officers is the reality of pollution control law, the book is concerned with the workings of regulatory control at field or street level, where officials make important screening decisions about the existence of 'problems' and the need for remedies. The work also addresses the conditions under which resort to the criminal process—a rather rare event in most forms of regulation—is regarded as fitting and desirable.

What is known about enforcement behaviour has been garnered almost entirely from work on the police, to the neglect of those countless numbers of individuals whose job is to enforce regulations punishable by the criminal law. Given the link between enforcement behaviour and the conduct to be controlled it is essential to examine enforcement directed to activity whose moral status is less than clear. Most studies of law enforcement reflect their ties with police behaviour in focusing on the 'problems' of non-enforcement or selective enforcement of legal rules. This book, however, takes a different approach and broadens the conception of enforcement to emphasize the question of *compliance*. Viewed from this perspective, highly selective use of the formal processes of law enforcement emerges more clearly as a valuable resource for enforcement staff and their agencies.

The book is, therefore, one about human values and human judgment. The central working concepts of the control of deviance by criminal law—'legality' and 'illegality'; 'misconduct' and 'compliance'; 'accidental', 'negligent', or 'deliberate' behaviour—are all widely-recognized notions which yield no obvious meaning when set in the context of the realities and complexities of the everyday workings of law. They are all matters for interpretation and discretion. The analysis, accordingly, is concerned with the *practical* expression of a legal mandate by regulatory agents: the adaptive processes by which pollution is defined, and an appropriate control response adopted. What tend to be taken for granted as 'pollution' and 'compliance' are the outcomes of organized, sometimes lengthy, social processes.

I intend a double meaning with the use of the word 'environment' in the title. While it is about the implementation of legal rules to protect the natural environment of rivers, flora, and fauna, the study is also about enforcement agents and organizations and the ways in which they respond to their social and political environments. My thesis is that law enforcement demands a moral as well as a legislative mandate. In pollution control work, enforcement is a consequence of moral rather than technical evaluations. Enforcement premissed on moral notions is recognizable to people, and a perceived identity of values shared by the enforcer, the regulated, and their interested publics grants the enforcement of regulation a more secure footing in an environment of ambivalence. Thus, although the legislature may equip enforcement agents with a strict liability law, it will rarely be exploited owing to the intervening lay conceptions of blameworthiness held by regulators and regulated alike. What is sanctionable is not rule-breaking as such, but rule-breaking which is deliberately or negligently done, or rule-breaking accompanied by an unco-operativeness which amounts to a symbolic assault upon the enforcer's and the agency's authority and legitimacy. The control strategies adopted do not directly reflect the values embodied in the legal mandate, but rather express a personal morality and a concern for individual and organizational self-preservation. In this context prosecution is almost always foreclosed, for everyday conceptions of morality intrude themselves at every stage of the enforcement process. From the point of view of legal policy, then, one of the implications of the analysis is that systems of control premissed on notions of strict liability will not facilitate enforcement. Another, more important, implication is that it is possible to conceive of the law being enforced even though the formal apparatus of prosecution is hardly ever used.

To study the enforcement of regulation is at once to explore some of the less familiar areas of social control, those 'shadowed places where administrative decisions are made' (Selznick, 1969: v), yet to discern more of the ironies of social control. By enquiring into the conduct of regulatory officials from the perspective of the more familiar law-enforcement behaviour of the police, the processes of rule enforcement and the place of the criminal sanction in the apparatus of social control may emerge in a clearer light.

The common theme in all of this is human judgment. Discretion is the stuff of the law. It is the means by which in everyday life legal mandates are interpreted and given purpose and form. It is the means by which judgments are made about the application, reach, and impact of the law. And it is the means by which the conflicting imperatives of consistency and diversity are reconciled. Through discretion, the law takes on substance and life.

Oxford, June 1981 K.O.H.

Acknowledgements

The work of staff in two agencies over a period of more than two years is described in this study. To conduct fieldwork over such an extended time obviously required the goodwill and patience of a large number of people. The customary undertakings about confidentiality and anonymity unfortunately forbid my identifying any of them by name (the names which do appear in the text are fictitious throughout). Here I simply record my gratitude to them. The degree of co-operation given and the many kindnesses shown me by so many officials in the field and at headquarters certainly do not support the suspicions of some scholars that public officials are unnecessarily guarded and unresponsive subjects of social research.

Environment and Enforcement is a product of the research programme at the SSRC Centre for Socio-Legal Studies at Oxford. I am grateful to the Social Science Research Council for funding the enquiry. The work reported in the book forms part of a larger project on the control of pollution, in which Yvonne Brittan, Max Hartwell, Anthony Ogus, and Genevra Richardson have, at various times, been involved. The Oxford Centre is an especially congenial research environment, and I appreciate the interest and support of my colleagues, particularly Max Atkinson, Donald Harris, and Doreen McBarnet. I am also grateful to Angela Palmer who had the tedious task of transcribing tape-recorded interviews and typing early drafts, and to Jennifer Dix who made an excellent job of the final typescript.

My wife Su has sacrficed her leisure time typing and checking seemingly innumerable rough drafts and keeping me (more or less) organized. I owe a special debt to her.

A (very) rough draft of the book was commented on by a number of American colleagues. For their helpful criticisms, and their patience in tackling a script occasionally rendered almost unreadable by an errant photocopier, I thank Gilbert Geis of the University of California at Irvine; Robert Kagan and Jerome Skolnick of the University of California at Berkeley (Skolnick also gave me some ideas for the title); Richard Lempert of the University of Michigan; Peter Manning of Michigan State University; Robert Rabin of Stanford University; Albert J. Reiss jun. of Yale University; and H. Laurence Ross and John M. Thomas of SUNY at Buffalo. I have not always followed their advice, hence I hasten to proclaim the usual absolution.

I. Regulating Pollution

These opening chapters describe and analyse some of the characteristic features in the enforcement of regulation. A distinction which is pursued throughout the book is drawn in Chapter 1 between compliance and sanctioning systems or strategies of enforcement. The chapter also presents a brief account of the formal organization of pollution control. The law itself makes general statements and formulates specific prohibitions, but it is up to administrative agencies to make policy and fill in the details. The regulatory agencies in pollution control themselves map out the boundaries and contours of the deviance they police by setting standards defining what is regarded as 'pollution'. That process, and some of its dilemmas, are explored in Chapter 2.

1. Introduction

I. Enforcing Law

Law may be enforced by compulsion and coercion, or by conciliation and compromise. In the enforcement of regulation, a distinct aversion is noticeable to sanctioning rule-breaking with punishment. Whether enforcement agents are concerned with air[1] or water[2] pollution control, consumer protection,[3] health and safety at work,[4] housing,[5] discrimination,[6] wage and price control,[7] or the many other areas[8] of social and economic life now considered to be the law's business, writers have observed a style of enforcement which seems to be predominantly conciliatory. Some, as a result, have complained that regulations are 'poorly enforced' and legislation is 'ineffective' (Gunningham, 1974:56; also Freeman and Haveman, 1973; Zwick and Benstock, 1971), or that 'a tradition of relatively weak enforcement prevails' (Bernstein, 1955:223). Implicit in this stance is a conception that criminal law enforcement is properly a matter of compulsion, which leads, inexorably it seems, to a conclusion that regulation has failed.

The enforcement behaviour of regulatory agencies has been explained most frequently in terms of 'capture' theory, according to which an agency is co-opted by those it seeks to regulate, incorporating and reflecting their concerns into its decision making in the interests of stability and self-preservation (e.g. Selznick, 1966). The shift is a subtle one in which 'the mores, attitudes, and thinking of those regulated come to prevail in the approach and thinking' of many regulatory officials (Bernstein, 1955:83).[9]

The *processes* of regulatory enforcement have not, however, been the subject of detailed enquiry. In this book the enforcement of regulation is analysed in terms of two major systems or strategies of enforcement which I shall call *compliance* and *sanctioning*.[10] I shall also talk of a *conciliatory*[11] style of enforcement as characteristic of compliance strategy, and a *penal* style as distinctive of sanctioning strategy. The terms 'conciliatory' and 'penal' are adopted from Black (1976) who discusses dominant styles of law which have counterparts in wider and more pervasive forms of social control. A conciliatory style is remedial, a method of 'social repair and maintenance, assistance for people in trouble', concerned with 'what is necessary to ameliorate a bad situation'. Penal control, on the other hand, 'prohibits certain conduct, and it enforces its prohibitions with punishment'. Its nature is accusatory, its outcomes binary: 'all or nothing—punishment or nothing' (Black, 1976:4).

Since the characteristics of sanctioning and compliance strategies are pervasive themes throughout the book, it would be as well to preface the analysis with a brief exploration of some of their general features.[12] I propose simply to map out broad characteristics, though the analytical language of contrast tends to suggest categorical qualities which are unintended. What may appear in their presentation as polar opposites are in reality shifting points on a continuum. Central to a sanctioning strategy is a concern for the application of punishment for breaking a rule and doing harm. Conformity with the law may be the consequence of this, but that is not the main issue. The formal machinery of law is crucial to this concern, and exacting a legal sanction by means of the legal process is a relatively routine matter. Enforcement agents who adopt a compliance strategy, however, are preoccupied with securing conformity to a rule or standard when confronted with a problem. Compliance strategy seeks to prevent a harm rather than punish an evil. Its conception of enforcement centres upon the attainment of the broad aims of legislation, rather than sanctioning its breach. Recourse to the legal process here is rare, a matter of last resort, since compliance strategy is concerned with repair and results, not retribution. And for compliance to be effected, some positive accomplishment is often required, rather than simply refraining from an act.

These differences are reflected in enforcement style. A penal style is accusatory and adversarial. Here enforcement is reflective: a matter of determining what harm was done, of detecting the law-breaker and fixing the appropriate sanction. The primary questions are whether a law has been broken, and whether an offender can be detected. If so, then the breach *deserves punishment*. In a compliance strategy, on the other hand, the style is conciliatory and relies upon bargaining to attain conformity. Enforcement here is prospective: a matter of responding to a problem and negotiating future conformity to standards which are often administratively determined. Since such standards are generally designed to prevent harm by accumulation, violations consist of rule-breaking which could lead to harm, as well as rule-breaking where actual harm is demonstrated (Reiss, 1980:30). This makes retribution inappropriate. If prevention of future misconduct occurs, it does so as a result of negotiation rather than the deterrence which (presumably) inhibits future rule-breaking in a sanctioning system. A standard which has not been attained in a compliance system *needs remedy*.

The emphasis given to detection and punishment in a sanctioning system is linked with a special concern for proof of violation (Reiss and Biderman, 1980:297). Decision outcomes tend to be binary and matters are ultimately settled by means of adjudication. As such, the process is visible, and a central role in adjudication is given to a stranger. In a compliance system, in contrast, there is much less concern for proving a violation took place; indeed widespread reliance on strict liability would make the question of proof relatively straightforward if matters ended up in court. Detection is important, however, but rather as a means of monitoring compliance and of enhancing prevention; indeed the commitment to repair of a potential source of harm produces a concern for the effectiveness of enforcement procedures in securing conformity. On the evidence of this study, the dominant conception of enforcement agents in a compliance system is a notion of efficiency: the attainment of a social goal at least cost to them and their work. Punishment is an unsatisfactory operational philosophy because it risks damage to the ultimate end of enforcement, and control of the case does not remain in their hands.

Decisions in a compliance system are graduated in character (Eisenberg, 1976), and though in rare cases matters are ultimately settled by adjudication, they are normally controlled by the parties themselves in private, intimate negotiations which rely on bargaining, not adjudication. Where enforcement relationships in sanctioning systems tend to be compressed and abrupt, compliance enforcement is marked by an extended, incremental approach. There are implications in all of this for what are regarded as indices of success for enforcement officials and agencies. Statistics of process, such as arrests and clearance rates are accustomed indices of organizational success in a sanctioning system. In a compliance system, however, statistics of impact are more likely to be employed to display the organization's effectiveness in repairing harm.

What prompts a sanctioning rather than a compliance response is not who does the law enforcement so much as the sort of behaviour which is subject to control. Where deviance[13] has a categorical, unproblematic quality, a penal response is triggered.[14] Sanctioning strategies in fact generally tend to be associated with incidents of deviance. The law-breaking is essentially a discrete activity, even though its consequences may be longer lasting. It is momentary, closed (Mileski, 1971). This isolated, bounded quality tends to make the deviance unpredictable and forecloses the possibility of a continuing relationship between enforcement agent and potential deviant. In general, the more unpredictable the distribution and location of deviance, the more a sanctioning response is likely (Reiss and Biderman, 1980). Where blame can be associated with the conduct, or where unco-operativeness (itself a blameworthy matter) on the part of the offender can be inferred, the response is also likely to be punitive.

Many forms of deviance, however, are continuing, repetitive, or episodic in character. They are states of affairs rather than acts: unfenced machinery, substandard housing, adulterated food, a dripping pipe, drunkenness or vagrancy, where that which gives offence is readily regarded as a 'problem'. Such deviance tends to be unbounded in time, providing for a continuing relationship between enforcer and potential deviant, and the matter of blameworthiness is often questionable. These sorts of rule-breaking are amenable to strategies of correction or control in a way that most forms of isolated crime cannot be. Where the deviance provides for the development of social relationships between rule-enforcer and rule-breaker, enforcement is directed towards compliance. Here enforcement is not the once-and-for-all response with the binary outcomes of a sanctioning system, but a serial, incremental, continuing process.[15] Compliance strategy may often involve highly elaborate and extended approaches, in which enforcement assumes a linear character.

Finally, victims of deviance also tend to hold different places in the two enforcement strategies. Personal harm prompts a sanctioning strategy. In a compliance system, in contrast, the victim is not necessarily a specific, readily-identifiable individual suffering obvious loss or harm, but a collective—often distant, diffuse, and indeterminate. Reiss and Biderman (1980:298) have put it well:

> Compliance systems appear to be concerned with victims in some aggregate rather than in a discrete or specific sense while penalty systems must deal with victims in the concrete since they constitute an element in their system of proof.

It is frequently difficult to speak, for example, of the 'victims' of water pollution. In some cases downstream users may have to close their intakes, and anglers may be appalled at the sight of dead fish. But when a pipe is discharging polluted water into a river which is largely an effluent channel, the only 'victim' may be the public, with the impairment to such amenity value as may remain.

II. Enforcing Regulation

Compliance and sanctioning strategies transcend institutional arrangements for law enforcement. It is tempting to regard the traditional criminal justice system, access to which is controlled by the familiar figure of the policeman, as typical of a sanctioning system of enforcement. But where policemen deal with states of affairs or episodic deviance such as vagrancy, prostitution, or mental disturbance, they too adopt a compliance strategy.[16] Similarly, regulatory enforcement—on the evidence of water pollution control work, at least—is not, by virtue of low prosecution rates, simply an example of a compliance system. *The enforcement of regulation incorporates elements of sanctioning strategy.* Indeed, some forms of regulatory rule-breaking are dealt with by a penal style of enforcement at the outset, and in other circumstances a compliance strategy will subsequently yield to a sanctioning approach if the struggle for conformity is lost. This suggests that law enforcement is more complex than some writers have assumed. Yet what *Environment and Enforcement* shows is the essential *similarity* in the behaviour of those who enforce legal rules.

The regulatory enforcement agent in routine cases adopts a compliance strategy which follows a serial pattern, a loosely structured but none the less organized process relying heavily upon negotiated conformity, with a gradual increase in pressure being applied to the unco-operative. In some cases the enforcement career may be rather compressed; in others, the business of enforcement may become very protracted, spanning months or years. When confronted with an instance of deviance, the field man's instincts are to stop the pollution, identify its source, and negotiate for preventive or remedial measures. A pollution does not provide an officer with the excitement, satisfaction, and possible prestige among colleagues enjoyed by policemen (Cain, 1973:32), but is a symbolic assault on 'his river', a serious blow to his careful tending of the environment to be healed as quickly as possible. Similarly, the officer's conception of an emergency is something demanding instant action to correct a problem rather than to apprehend a violator.

The place of the formal legal process—prosecution—is transformed in this compliance system from the public, occasionally dramatic, but quite conventional response in a sanctioning system to a rarity reserved for a small, highly selected number of cases. In relation to the instances of prosecutable deviance which come to light, few pollutions[17] even result in the initiation of the formal process by the taking of a statutory sample; rarely do they end up in court with conviction, sanction, and the attendant publicity in the local newspaper (see ch. 9, s.i). And the drama is hardly ever played out in the Crown Court.[18] When prosecution for pollution does take place, it is in the magistrates' court, where it shares the stage with licensing applications and speeding motorists. The threat of prosecution lurks, however, in the background of private negotiations, a threat to be unveiled in the face of unco-operativeness or intransigence. Where enforcement in a sanctioning system is occasionally dramatic, securing compliance with regulation has little potential for drama. It tends to be uneventful, its often prolonged sequence of steps suggesting enforcement by attrition. The control of water pollution, for example, like other forms of regulatory behaviour, is an unobtrusive activity. In contrast with the readily-recognizable policeman on the street, pollution control is conducted by agents who, bearing no insignia of office, are for the most part obscure and anonymous. The work itself consists substantially of personal transactions between the polluter[19] and field staff. Their privacy and low visibility may be exploited by enforcers who, as a strategic consideration to secure compliance, can hold out the prospect of the ultimate sanction for deviance—the public display of the polluter as a defendant in court. In this context, prosecution itself—and not the punishment imposed by the court—becomes for the regulatory agency the penalty for failure to comply. However, the enforcement agent in a compliance system regards prosecution as a sign of failure, where in a sacntioning system it becomes visible evidence that he has done his job. Compliance strategy allows justice to be done in the process of negotiating conformity, whereas in a sanctioning system justice is done (so far as the enforcer is concerned) when someone is let off (Reiss and Bordua, 1967:37-8).

One of the themes of this book is that strategies of control employed in the enforcement of pollution regulations are shaped by features inherent in the nature of regulation itself. In particular, regulatory control is characterized by an ambivalence[20] which has both political and moral dimensions.

Regulatory agencies must operate in a political environment, for regulation is intended to preserve the sometimes fragile balance between the interests of economic activity on the one hand and the public welfare on the other. Agencies are extremely sensitive to their political environment.[21] They find themselves operating between two broad publics or constituencies with competing views about the proper realm of government in regulating the economy. These constituencies are a reflection of ideological differences and represent opposing positions on the fundamental political dilemma of regulation: the extent to which economic restraint by the imposition of legal rules is justifiable (Edelman, 1964:22 ff.; Freedman, 1975; Kagan, 1978:9 ff.; Murphy, 1961). These constituencies are broader and more diffuse than mere pressure groups, though such groups may comprise their most visible and vocal components. They are, rather, constellations of political ideas, currents of opinion, with interests in agency activity. From time to time they surface to make regulation a topic of political debate. These interested, shifting publics are the agencies' significant audiences.

I shall caricature the constituencies in pollution control as 'environmentalist' and 'business'.[22] Those sympathetic to the environmentalist position accept the need for substantial expenditure on water pollution control. They urge a fuller, more activist policy of enforcement, and advocate wider restraint on economic pursuits, in the interest of minimizing the harms which flow from unregulated activity, even though this may impose substantial costs and restraints upon productive enterprise. Agency enforcement is often criticized from this perspective as scant and ineffective, with the agencies captured by the interests they seek to control. Legal sanctions are regarded as impotent.

The business constituency, on the other hand, views much regulation as an unjustifiable intrusion by the State.[23] It proclaims the burdens of pollution control regulation, displaying industry as suffocating under costly yet trivial constraints. At the same time it plays down the dangers of pollution. Those sympathetic to industrial and agricultural interests seek a 'hands off' policy, with few impediments to productivity. The business constituency not only wants to be as little burdened as possible with the costs of complying with regulation, it is also critical of what it sees as inordinate amounts of money being spent on pollution control by bloated, publicly-funded organizations.

The existence of these two adversaries reflecting incompatible interests poses practical problems for the water authorities, for while they have been established ostensibly to advance environmentalist values, and serve as a concrete expression of the belief that intervention is proper and necessary (for which they have been equipped with the criminal law), they find that their legitimacy as enforcement agencies is sometimes questionable. Enforcing rules when caught between opposing public constituencies causes difficulties for agencies which must work within the framework of an ostensibly activist legal mandate. Organizational self-preservation, as this book shows, makes it imperative for them 'to manufacture the appearance of activity ... the symbolic reality of impact, the fiction of real power ...'.[24] The resulting sense of vulnerability is the more acute because regulation implies a degree of tolerance about the activity causing concern, rather than its elimination. From a social policy point of view *the issue is not whether to allow pollution, but how much pollution to allow* (see Ackerman *et al.*, 1974). Pollution control work, then, is typical of the many areas of social control characterized by goals of *regulation* rather than *repression*. Regulation is a practical compromise between the benefits and harms of unfettered economic activity. 'The real source of deviation in such areas', writes Lemert (1972:55), 'is not necessarily change in the behavior of the subjects of regulation, but may be the imposition of new rules which define existing behavior, or behavior consistent with older norms, as now deviant. The object of so defining the behavior is to produce change, not to repress it.' *As soon as we talk of 'regulation' rather than repression, we admit the necessity of discretionary enforcement and open the way to controversy.*

The moral dimension of the ambivalence surrounding regulatory control is most clearly exposed by regulatory rule-breaking. There is a reluctance, prima facie, to regard a breach of regulation as morally reprehensible, since the conduct addressed is widely regarded as 'morally neutral' (Kadish, 1963; also Ball and Friedman, 1965; Fuller, 1942; Yoder, 1978), in contrast with those behaviours which are the stuff of traditional criminal law. This is recognized in the familiar distinction in criminal law between *mala in se* and *mala prohibita.* Prohibitions are nevertheless crimes because, in Matza's words, 'they elicit authorized state intervention, but they are different from other crimes in failing to self-evidently warrant intervention' (Matza, 1964:161). Those crimes that 'self-evidently warrant state intervention' I shall refer to throughout as 'traditional crimes'.[25]

Yet to describe regulatory deviance as 'morally neutral' misses the point. 'Morally problematic' might be more apt, for later chapters will argue that all rule-breaking is morally evaluated, with profound implications for enforcement. Regarded in the abstract, pollution offences may not be treated as 'real crimes'. Most people, indeed, are unwilling to talk of 'crime' when they discuss breach of pollution regulations: this sort of language is considered appropriate only where clearly blameworthy conduct exists—where there is a calculated breach of regulation, or where the polluting substance concerned is widely known to be dangerous and there was carelessness or recklessness in handling it. But when clothed with notions of gravity or blameworthiness—and setting is an inevitable component in the evaluation of conduct—any hints of moral neutrality are abandoned.

Moral ambivalence is probably associated with a number of other features which distinguish regulatory misconduct from breaches of the traditional code. One is that regulation seeks to control economic life, yet it is this which is recognized as responsible for the material well-being of the community. Another is the relative recency with which new values, untouched by the aura of the sacred, have been invented and proscribed (Sutherland, 1945; but cf. Walsh and Schram, 1980). Pollution, for example, became a problem suitable for regulation by the criminal law as a result of nineteenth-century industrialization and urbanization (Brenner, 1974). In effect the law has encroached upon vast areas of activity. A broad array of newly-defined prohibitions, often serving several conflicting social goals, has been grafted onto a criminal law concerned with traditional crimes—such taken-for-granted unacceptable behaviours as murder, rape, robbery, arson, and theft (see Diver, 1980; Thomas, 1980). Furthermore, the definitions of these latter forms of deviance, the prototypical subject matter of police enforcement work, are located in legislation and its attendant case law. In other words, the attributes of traditional crime are defined for the enforcement agency. They have a general audience. Regulatory agencies, however, are typically vested with a broad legal mandate which gives them discretion to establish standards marking out legally acceptable and unacceptable behaviour (see Lowi, 1979). This presumably does not make the prohibitions any less morally problematic.

In Britain, ambivalence about regulatory rule-breaking has been displayed in the levels of sanction provided in the criminal law (see Carson, 1980b). It is as if the legislature wishes both to compromise the criminalizing effect of conviction and to mitigate the impact of a prosecution facilitated by strict liability (cf. Friedman, 1967; Paulus, 1974). While the penalties for water pollution offences are in theory more severe than most field staff believe them to be, they remain frail in comparison with those available to sanction traditional crime.[26] Though regulatory law is in general framed to penalize leniently, complaints are sometimes made (particularly by enforcement agents) that the levels of penalty imposed are themselves derisory. One of the two pollution control agencies studied reported, for instance, that in a recent twelve month period the average fine levied for pollution offences was £49 (unpubl. agency document).[27]

The harms which are associated with breaches of regulation also encourage a degree of moral ambivalence. Harms which are the consequence of most traditional crimes are normally immediate and noticeable, and establishing a direct link between act and consequences is a relatively simple matter which facilitates enforcement. While harm may sometimes be direct and visible in the regulatory arena, it may also be vague and amorphous, gaining in severity by accumulation. Indeed, although regulatory offences in the abstract may be regarded as of minor consequence they may in certain cases have drastic results. Similarly, the victims of pollution may be readily apparent in some cases, but in many others they may be dispersed and diffuse. Some victims of regulatory deviance may be made remote by time, as in the case of air pollution, radiation, or some forms of occupational disease. Here, each individual act or continuing infraction may not amount to a significant harm of itself. It is in the aggregate that the damage may assume serious proportions, hence the emphasis given to preventive work in compliance systems. The victims of this kind of deviance may be unaware in the absence of a perceptible physical threat that they are being victimized. In this connection it is difficult, sometimes impossible, in pollution control work to establish a satisfactory causal link between event and harm. Indeed, it is not even known whether some substances which contaminate water are harmful or not. It is predictable that impetus for regulatory reform or pressure for crackdowns generated by moral entrepreneurs (Becker, 1963) often comes about only when victims become visible, for it is the impact of deviance which contributes to a judgment of its gravity (Schrager and Short, 1980). It seems to be more difficult for human beings to apprehend the gravity of harms which take several years to manifest themselves.

Ambivalence poses the crucial problem of enforcement for regulatory agencies and their field staffs, because *their authority is not secured on a perceived moral and political consensus about the ills they seek to control.* The police, in comparison, enjoy a relatively secure moral mandate. In pollution control work, however, there is none of the sacredness of the policing of the traditional code (Lemert, 1972; Manning, 1977), and it is more difficult to dramatize the threat of pollution than to portray the symbolic assaults on the community from criminals, addicts, vandals, and other sinister figures on the fringes of the moral order (Manning, 1980), notwithstanding the missionary zeal of some proponents of regulatory reform in the USA. Regulatory deviance rarely possesses the emotive properties of many traditional crimes. The latter also more regularly invite the attention of the press and therefore the organized resentment of the public.[28] In the traditional code there is rarely any 'good' reason for a breach. In regulation, however, there is a myriad of 'plausible' grounds for non-compliance or partial compliance. The lack of a moral mandate threatens the regulatory agency's legitimacy as an enforcement authority. Instead, agencies find they must tread a tricky path between the competing claims of state regulation and free enterprise. As this book shows, the ambivalence, tensions, and dilemmas prompted by the opposing interests in regulation are continuously worked out in the day to day enforcement decisions of field officers.

Three other matters central to regulatory enforcement deserve brief attention. First is the question of liability. The relatively recent introduction of regulatory prohibitions has been associated with an extension of the law's conception of criminality to embrace a notion of social harm in addition to one of individual guilt, characteristic of traditional crime (cf. Sayre, 1933). This shift has been reflected in a desire to protect the public from the unanticipated consequences of everyday applications of science and technology and accompanied by the growing role of organizations as actors in society. The sensitivity in the law to collective interests and the problems posed where the law has encroached upon complex technical behaviour have provoked concern about the difficulties of proving intent, particularly where organizations are involved. These in turn have raised difficulties of evidence and proof in law which legislatures have attempted to circumvent by rejecting the traditional prerequisites of some degree of mental intent on the part of the deviant actor—the concept of *mens rea*—in favour of the ostensibly more easily managed notion of strict liability, permitting (in theory) legal enforceability of a prohibition without regard to the blameworthiness of the actor (*Harvard Law Review*, 1979; Jacobs, 1971; Paulus, 1974). Here, again, we see the focus upon ends characteristic of compliance systems, for 'the achievement of the regulatory purpose is seen as the all-encompassing justification' (Allen, 1977:756). In traditional criminal law much enforcement activity is given to establishing the existence of *mens rea*, creating the conditions for the application of legally defined blameworthiness, since it is this which is a prerequisite for the imposition of punishment. *Mens rea* is not normally an issue in the case of pollution of watercourses when cause is to be established. The law speaks of causing or knowingly permitting pollution, and the concept of cause has been strictly construed.[29]

Secondly, mirroring the extension of the law's concern to notions of social harm has been a widening of the conceptions of deviant and enforcer to include not only individual actors but also organizations. Theorizing and debate about law-breaking and law enforcement have been premissed on a model of the criminal actor as individual and on a conception of enforcement as practised by a uniformed and public police. Regulatory control, however, is concerned for the most part with organizational deviance, and with many activities in which compliance does not reside simply in refraining from an act, but in positively doing something to remedy a state of affairs. Much of a regulatory enforcement agent's behaviour is moulded by the fact that he is confronted with organizational activity, where the policeman is typically concerned with individuals or groups. This dimension also has implications for deterrence theory: since organizations comprise many individuals of varying occupations, degrees of organizational responsibility, and access to deviant conduct, who, precisely, is being deterred, and how are they deterred, when a company is convicted?

Thirdly, regulatory agencies are specialist enforcement authorities, having a finite domain of control, consisting of particular behaviour and involving only a segment of the population at large, defined by the kind of economic activity subject to regulation. Such a focused population is continuously open to control, and the population of deviants is potentially knowable (Reiss and Biderman, 1980:276) giving rise to an appreciation of potential violators as possessing unique identities. The intimacy of the enforcement relationship is a characteristic of compliance systems where the control of a distinct population of potential violators is directed to 'organizations or persons in organized activities [rather] than to individuals apart from them' (Reiss, 1980:31). Such a personalized system of enforcement furthers the commitment to effecting remedy, inherent in compliance strategy. Police enforcement of the traditional code, on the other hand, is in theory not a matter for a mere segment of the population, and except where the police deal with persistent or 'problem' offenders, their target population is one made up of strangers,[30] given to unpredictable deviance.

Left unremarked in the discussion so far is the irony of the place of criminal punishment in a system which emphasizes conciliatory control based upon compliance. This irony prompts the questions which guided the research and which form the central problem addressed by this book: *since those who are subject to regulation have good economic reason not to comply, how is compliance secured, given the frailty of the criminal sanction and its virtual disuse? How is control effected where the law seeks to remedy a state of affairs? What is the place of the formal legal process in regulatory enforcement?*

I take a crude conception of compliance as a starting-point, since it is clear that, in factory or farm, money has been spent in response to the existence of water pollution regulation.[31] The simple physical evidence of at least a measure of compliance is widespread. Anyone who walks round industrial premises will find visible sign of pollution treatment facilities installed and maintained, often at considerable expense.[32] The efforts farmers have made to control their silage liquors or cowshed washings are equally evident. How have pollution control officers managed it? How is regulatory law translated into regulatory activity?

III. A Research Perspective

The present essay is a piece of interpretative sociology in the tradition of the societal reaction school (e.g. Becker, 1963; 1964) and the dramaturgical approach of Goffman (e.g. 1959; 1961; 1963: 1967; 1970; 1971). It makes particular use of the work of scholars who have conducted ethnographic analyses of discretionary behaviour in legal settings (e.g. Cicourel, 1968; Emerson, 1969; Manning, 1977; Reiss, 1971; Ross, 1970; Skolnick, 1966; Sudnow, 1965). An account of the research methods employed is presented in the Appendix.

Regulation and its enforcement are treated as social processes. What might be regarded as the 'problem' of pollution resides not simply in the physical environment but in social behaviour. A 'pollution' is a creation of human judgment. Emphasis is given to the fluid, adaptive, but essentially patterned character of enforcement behaviour. Legal rules are mediated by the intervening interpretative processes informing the actions of enforcement agents which give substance to the vague aspirations of statute. Similarly, evaluations of polluters are treated as open to the application of various definitions for the purpose of enforcing regulations. Thus, in presenting in Part II of the book a descriptive analysis of the routine work of field officers, the working definitions they observe in doing the job are stressed. After all, pollution control is what field officers do every day in factories and farms (cf. Kaufman, 1960:65).

The ways in which events and acts come to have meanings attached to them which serve as the bases for action or inaction by regulatory agencies, and the negotiating tactics employed in securing compliance are central topics dealt with in Part III. A focus of the analysis here is upon the characterizations generated by enforcement agents of 'polluter' and 'pollution' which inform the sequence of judgments comprising the agency's response. To focus upon the processes of characterization is to draw attention to the definition of act or event, and interaction between potential deviant and enforcement agent. Such interaction has implications for the future behaviour of field staff, for enforcement is organized to reflect the ways in which polluters define their behaviour and the 'pollution'. Control, in short, is regarded as a reciprocal relationship. The enforcement relationship is accordingly presented as negotiated and symbiotic in character, since social control encounters, 'those where reaction to deviance is central, involve degrees of mutual dependency, thus all rule enforcement is a matter of negotiation' (Manning, 1977:247, italics omitted). In some circumstances interaction may prompt enforcement activity substantially independent of the definitions of the seriousness of the problem to be remedied. The interpretative work of field staff, particularly the procedures by which they graft meaning onto the behaviour of polluters for the purpose of categorizing them for future decisions about enforcement, is, then, a central theme.

The perspective presented is one from field level, though it is one which recognizes that an agency's legal mandate is refracted by organizational constraints and demands. Problems of order, control and freedom are, indeed, increasingly caught up in organizational processes (Manning, 1980:10). My intention is to display pollution control work as individual and organizational activity premissed upon the appreciation of symbol and common-sense theorizing about behaviour. In a bureaucracy such as a pollution control agency, the organizing principle is administrative efficiency—'an orientation to the expeditious attainment of the given objectives' (Blau, 1963:264)—which reaches down to the field officer in the form of a number of imperatives about getting the job done in certain ways that have profound implications for his exercise of discretion. The impact of bureaucratic strategies of control imposed on the enforcement job in a setting characterized by a high degree of personal autonomy is explored in Part II.

The analysis also pursues some of the contrasts between sanctioning and compliance systems of enforcement, using the now substantial literature about the policeman on the street[33] to throw the behaviour of the pollution control field staff into bolder relief. Anglo-American sociology of law has been imprinted with the emphasis given to the study of law enforcement and criminal justice systems in handling traditional crime and, despite the interest of a few sociologists of deviance in some marginal forms of deviant conduct, our present conceptions of criminal behaviour are heavily influenced by this emphasis. The literature, in fact, has left regulatory misconduct, deviant organizational behaviour and organizations as the objects of enforcement activity virtually untouched, with the celebrated exceptions of Sutherland (1940; 1949), Clinard (1952, 1979) and Geis (1967, 1968).[34] But even among these writers the predominant conception of organizational crime is of individual traditional crime, in the form of the various kinds of dishonesty committed by corporations, rather than one of the violation of the host of criminally-enforceable regulations to which organizations are subject. This is unfortunate. To look at the enforcement of regulation may help refine some of the present conceptions of enforcement behaviour and organizational deviance, while regard for the enforcement of prohibitions may sharpen perspectives on the nature of traditional crime.

IV. The Formal Organization Of Pollution Control

The essence of regulation is a conception that law is the means by which some notion of the public good is to take precedence over narrow economic interests. This is not a particularly novel view in law. What is of relatively recent origin, however, is the creation of bureaucracies equipped with legal sanctions to regulate economic life. Specially created to control a particular segment of economic activity, regulatory agencies in general enjoy broad powers to set and enforce standards of conduct.

This book reports the control of one such activity, namely pollution discharged directly to watercourses, excluding trade effluents which are disposed of to foul sewer for purification at sewage treatment works.[35] It should be remembered, however, that the great majority[36] of inland sewage works themselves ultimately discharge their treated effluents to open watercourses, subject to more or less the same kinds of constraint as other dischargers.

Pollution control is administered in England and Wales by Regional Water Authorities which were created by the Water Act, 1973. These authorities are presided over by the National Water Council, set up at the same time to be a central co-ordinating body to give advice on major matters of policy (*Times*, 1978). The task of the authorities is to manage all water services within a geographical area determined by river catchments. The ten authorities acquired control of water services from a variety of other bodies including river authorities and local authorities, and are now responsible for all matters connected with the use and disposal of water: water conservation and supply; land drainage; amenity, recreation, and fisheries; and pollution control. Each water authority has its own form of organization, administrative control, and territorial jurisdiction. The responsibility for detecting pollution and enforcing regulations in the two water authorities studied rests with a policing section or inspectorate. The authorities must therefore police themselves, a problem the more poignant since their own sewage works are often the worst polluters.

The authorities' legislative mandate, intended to give organizational activity a sense of purpose and direction, is framed in general terms and confers wide discretion as to the formulation of policy and its implementation. The Water Act, 1973, speaks of a national policy for water, 'the restoration and maintenance of the wholesomeness of rivers', and 'the enhancement and preservation of amenity' (s. 1). It goes on to specify a duty to take action 'necessary or expedient ... for the purpose of conserving, redistributing or otherwise augmenting water resources' and 'securing proper use of water resources' (s. 10). There is no more detail than this. The principal offence,[37] located at the time of field work[38] in the Rivers (Prevention of Pollution) Act 1951, s. 2, is committed if a person 'causes or knowingly permits to enter a stream any poisonous, noxious or polluting matter ...'. The offence is in the main one of strict liability; indeed in the leading case of *Alphacell* v. *Woodward* (1972), the House of Lords endorsed a strict construction of the word 'causes'.[39] The 1951 Act also provides for personal as well as corporate liability.

The penalties in force during the research for a breach of s. 2 of the 1951 Act were a maximum fine of £200 for conviction on indictment, £100 on summary conviction. However, if 'the offence was substantially a repetition or continuation of an earlier offence' and followed conviction for the earlier offence, the penalties were increased substantially. On indictment a prison sentence of not more than six months was possible together with a fine;[40] on summary conviction the maximum was three months.[41] Stiffer penalties for the principal offence embodied in the 1974 Act were awaiting implementation during fieldwork: these allow for a fine of £400 or three months imprisonment, or both, on summary conviction, and for 'a fine' (without maximum) or two years imprisonment, or both, on conviction on indictment. The legislation does not provide other forms of sanction—though agencies are prepared to employ sanctions of their own as part of the compliance process (ch. 7).

In this study I shall refer for convenience and anonymity to 'pollution control agencies'. It must be remembered, however, that what are described as 'agencies' are sections (albeit independent sections) of a much larger authority. Also for reasons of anonymity, I adopt neutral terms to define the pollution control district officers, referring to them as 'field officers' or 'field men'.[42] Their immediate superiors are described as 'area supervisors' or 'area men'. Headquarters staff are referred to as 'senior staff'. The research was conducted in two water authorities which when necessary I shall refer to as the 'northern' and 'southern' authorities—adjectives simply chosen in the interests of anonymity.

In both agencies pollution control is administered from headquarters by an independent department.[43] The northern agency's territory is very mixed. Many parts of it are extremely rural with farming the predominant industry; watercourses here are maintained to a high standard of cleanliness and relied upon heavily as potable supply for the many industrial cities elsewhere in the authority's jurisdiction. Headquarters is located in a major conurbation where there is a wide variety of manufacturing industry. Some of these works are small, some exceptionally large, sometimes employing thousands of people on one site. Many of these are extremely old; many are run down. Rivers in this area are mainly comprised of domestic and industrial effluent, and many have been fishless in living memory. There is plenty of heavy industry in other cities in the authority's areas, including coal mining and steel making. Leachate and contaminated ground water from old industrial sites are major problems in urban areas where streams are small. The authority's sewage works vary in size from those producing many millions of gallons of treated effluent a day to those serving a few houses in a rural area where the discharge may be the merest trickle. In such a mixed territory it is no surprise to find that the quality of watercourses is often strikingly disparate. Many are little more than effluent channels, while some, by an accident of catchment, contain water fit for potable supply. A familiar feature in the northern agency's territory is the watercourse with headwaters of good quality which becomes heavily polluted downstream by the effluent from a sewage works serving a large town, or the discharges from one or two major industries. In general, in the authority's industrialized areas the emphasis in pollution control is on cleaning up heavily polluted rivers.

Pollution control work in the southern area has had, in contrast with the northern, a relatively long history and tends as a result to emphasize the maintenance of already existing acceptable standards of water quality. The southern authority area is dominated by a heavily populated conurbation. There is a good deal of industry associated with the city, but for the most part it is not heavy industry. Pollution control problems arise from the number and variety of industrial and manufacturing processes employed in industries set in an entirely urban environment, since domestic sources, with few exceptions, are efficiently treated in a small number of large sewage works. Elsewhere in the southern authority areas there are a number of towns with light industry, much of which is modern, with effluents discharged to foul sewer rather than direct to watercourses. In the headwaters of most rivers there are large tracts of agricultural land and a virtual absence of industrial discharges. This provides for a supply of very good quality water to the main river, which itself is used as potable supply. Many discharges, both domestic and industrial, are made into the tidal section of the main river.

Within each agency, the formal organization of pollution control is moulded by topography with river catchments determining the districts of individual jurisdiction. The district is the basic geographical unit of pollution control and usually contains one or two river or stream systems.

In the northern agency, about half a dozen districts comprise a pollution control area. Each district is policed by a field officer responsible to an area supervisor. Each area has an administrative headquarters, usually located in one of the more important towns near the area's geographical centre.[44] In addition to the half dozen or so field officers patrolling the districts, some of the areas in the northern authority have provision for a small number of additional posts. There may be one or two senior field officers whose job is partly to give administrative assistance to the area supervisor, partly to carry out routine pollution control work in their own districts. These senior officers supervise the activities of the one or two assistant field officers also found in most areas. This junior rank was introduced by the agency during the course of the research in an attempt to establish a training grade for field staff and as a means of relieving district officers of some of the simpler but time-consuming tasks of routine river sampling. Field assistants, unlike the district men, are not responsible for any geographical part of an area.

Organization in the southern agency is less centralized. It is subdivided into three geographical and administrative regions which are independent except at the highest level. Fieldwork was conducted in two of the three regions, each of which has its own particular administrative structure. In one region, a heavily urbanized area, the arrangements resemble the northern agency's internal structure, with an area supervisor directing a number of field officers each tending a discrete geographical district. In the other region (a mixed area containing substantial rural zones) there are, in addition to the field officers, a number of senior and junior posts. Here tasks are less clearly differentiated according to rank; men in all three grades would, for example, conduct routine river sampling.

In both agencies there are in effect four senior administrative positions with direct responsibility for pollution control work. The most junior of these acts as a regional supervisor who manages the activities of area staff. The three senior officials are generally involved in policy and planning issues and the management and co-ordination of pollution control operations. The directors and deputy directors are key figures in the formulation of pollution control policy, and in decision making about the use of law in pollution control. They are advised by and responsible to a committee of members of the water authority.

Both authorities have legal departments whose staff are, from time to time, called upon to offer expert advice where senior officials are dealing with pollution control matters with a legal content. Expert advice is most frequently needed when senior staff are considering mounting a prosecution (ch. 9) or when there are difficulties in connection with the issuance of a consent (ch. 2).

V. Postscript

Modern law, Lowi has observed, 'has become a series of instructions to administrators rather than a series of commands to citizens' (1979:106, italics omitted). Water pollution control shares with other forms of regulation a broad legislative mandate and an unspecified task of implementation. A characteristically legal mode of social control—the criminal sanction for misconduct—is made available to govern a social problem with technological and economic dimensions. The legislature creates a rather abstract mandate and an agency to implement it, while only defining explicitly the offences which give rise to prosecution. And even then, the content and boundaries of the offensive behaviour are matters of administrative discretion. Ironically, the power to prosecute is rarely used. Legislative mandate and legal offence are linked by an unguided and pervasive administrative discretion. It is as if the creation of a regulatory bureaucracy carries with it implicit powers to achieve agency goals. The power to negotiate, which is central to compliance systems, is one which exists by implication only but one made possible because the law is not made concrete or specific. Regulatory law is permeated with uncertainty.

2. Setting Standards

I. Issues

The boundaries of regulatory deviance are drawn by administrative agencies: pollution, in other words, is an administrative creation. The broad legal mandate of the agencies about water pollution control is transformed into policy by senior officials and given practical expression in the setting of pollution standards.[1] Standards ('consents') are licences to discharge polluting matter. Pollution is in effect qualitatively and quantitatively controlled by the water authorities since standards are administratively negotiated. Not only, then, do the agencies possess power to enforce the law, they actually determine the reach of the law, for (in contrast with the police) they exercise a real legislative authority, enjoying broad discretion to define what makes 'pollution'. In this sense the water authorities create pollution, as Becker might say (1963:9), by making the rules whose infraction constitutes pollution. Definition and enforcement, a dual authority, are reciprocally related, a theme to be explored in this chapter.

For agency policy-makers the consent is an important tool expressing political and economic judgments about water quality. Its central purpose, as the head of one the agencies put it, is 'to produce a river which is suitable for the uses which are needed downstream —providing a potable water supply, for fisheries, just amenity. We might even decide we just want an effluent channel. ... The consents are geared to produce the quality water you want in the river for the use you want downstream.' For the water authorities in general the practical significance of consents is that many rivers are now substantially comprised of effluent already discharged subject to consent. One of the southern authority's main potable supply rivers, for example, consists in dry weather of more than 50 per cent effluent; and another river flowing through a major city in the northern authority is over 90 per cent effluent.

Consents to pollute come in the form of emission or effluent standards which prescribe the temperature, amount, and kind of polluting matter which may be discharged from a particular source. Dischargers are thus permitted to pollute by the water authorities—

but only (in theory) up to those levels set out in the discharger's consent document. A consent, however, is a movable threshold. Once fixed it may subsequently be reviewed and modified by the agency in consultation with the discharger. The pollution standards in a consent are defined locally by each water authority and are specific in application, with each consent negotiated on an *ad hoc* basis. Standards are expressed quantitatively and each one is formulated not to contemplate exceptional circumstances, instead addressing pollution without regard to mitigating features, such as accident. Simply by selecting those substances in impure water which are to be held potentially 'polluting' (pollution parameters) and the point at which such contamination is to be regarded as 'polluting' (pollution limits), together with temperature and volume restrictions, the agency establishes theoretically enforceable boundaries,[2] exercising, in other words, power to control the *potential level* of pollution. And since the water authorities are enforcement agencies they also control to a very great degree what may seem to be some sort of '*real level*' of pollution which comes to light in amount and kind—those events or incidents which are detected and processed, becoming statistics of non-compliance or even, in some rare cases, of prosecutions in annual reports.

In setting standards the water authorities, like other regulatory agencies, are confronted by two crucial problems. First, pollution control means cost. Since the cost of control usually returns few, if any, financial benefits to the discharger and has to be met by increased prices, dischargers regard pollution control costs as a burden to be avoided wherever possible. There is no simple relationship between control and costs. To make a significant impact upon a bad discharge is not necessarily very expensive. But to make even a minor improvement to effluent which already consists of relatively good quality water may involve the discharger in very heavy expenditure indeed (see Kneese, 1973). The power to define and enforce consents is ultimately a power to put people out of business, to deter the introduction of new industry or to drive away going concerns.

The second, related, problem in standard setting is one of the distribution of costs. A tension between officials' perceptions of equity and efficacy is, for them, a familiar dilemma. The conflict is between an essentially moral stance which prizes consistency and uniformity in the application of standards (similar discharges should be similarly controlled in degree and kind) and a utilitarian approach concerned with effect on water quality, regardless (again, in theory, at least) of the means of the discharger or the demands fortuitously being placed unequally on similar dischargers (see generally Ackerman *et al*, 1974). '... [T]he load of polluting matter discharged should be controlled at all times to exactly the level that a river would accept without harming river uses or the environment' (unpubl. agency document). This is a recurrent dilemma in pollution control since efficacy insists that standards be tied to the setting of a discharge and settings vary enormously.

The weight to be given to the normally competing values of equity and efficacy is a persistent problem both in setting or reviewing standards and in enforcing them.[3] On one level the dilemma is between standards of general and of individualized application. On another the conflict is sometimes portrayed as between older, established ways of handling pollution control and newer, more 'scientific' approaches in which environmental impact in its widest sense is the primary concern. The utilitarianism of the scientist tends to view the principle of equity with a certain disdain because of its potential inefficiency: 'Equitability can be the biggest millstone,' said a senior officer with a special interest in consents. 'Often it goes against technical judgments. When it comes into play, some of the ultimate answers are not the best ones.' In terms of formal policy, a shift in recent years has seen equity ostensibly yield place to efficacy. In practice this should mean that the design of consents has been influenced less by conceptions about the similarity of discharges than by a scientific analysis of the impact a particular discharge may have on a particular watercourse and the polluting load that that watercourse can accept, according to agency plans for river quality.

Senior officials are firmly committed to the utilitarian view: 'treating discharges alike in terms of ... the quality of the discharge, as distinct from the effect on the environment', said one, 'is absolute nonsense.' Yet senior staff and field men alike are aware of the fact that most dischargers are likely to be more sympathetic to an approach to standard setting and enforcement based on equity. Apart from the moral preference that like should be treated alike, dischargers' commercial instincts argue that no manufacturer or producer should be placed in a more favourable position than his competitors. The practical task, then, for a water authority setting or modifying a consent based on criteria relating to the particular watercourse becomes one of persuading a discharger disadvantaged by some criterion which makes sense in utilitarian—but not moral—

terms of the force of the agency's position. In some cases this is apparently not as arduous a task as it may seem, simply because dischargers find it extremely difficult to portray themselves as 'similar' to their rivals on more than a very few criteria.

An individualized approach based on a concept of efficacy can be sold to dischargers because it is 'scientific'. Equity can be played down:

> 'I don't know if it should be that important if everything we do is based on good sound scientific principles, because if people perhaps in the same industry are situated in different places in the estuary, y'know, if one was to point the finger at the other and say "But you allow him to discharge such and such and you only let us do this", then we should be able to turn round and say, "Ah yes, but you're discharging in a different place and the river quality in this different place needs different treatment." ... I think you can only treat them similarly if all other things are equal, if they're discharging into the same sort of watercourse in the same sort of position. Because obviously if you had one particular factory on the tideway and another producing an identical product and identical effluent in one of the tributaries, you couldn't treat them the same because they're patently different cases.'

On this 'scientific' view of standard setting, the design of consents to discharge polluted water is to be based on a dispassionate consideration of the amount and kind of pollution load any watercourse can bear. The capacity of a watercourse and the tolerable limits of polluting effluent are calculated according to a mathematical model, which ideally should be applied routinely and objectively to all dischargers regardless of their means, the costs of treatment, their prior efforts, or the demands made of their competitors.

Field men, however, are constantly reminded that industrialists and farmers work on a principle of equity: they can readily discover the standards which their competitors must observe and may complain if they are being handicapped. But the agencies prefer to avoid complaints wherever possible, and this encourages an administrative inclination for equity of treatment, even though 'scientific' judgment may dictate otherwise. Negotiating about standards, especially when there is some disparity between apparently similar dischargers, though an infrequent event, can be one of the field man's trickiest tasks.

The picture of an agency which dispassionately administers scientifically-designed standards is blurred further by organizational practices. Later chapters will show that in pollution control work, standards are by no means treated as absolute proscriptions inexorably enforced. The agencies display a sometimes considerable flexibility both in the standards set and in the enforcement policy adopted, in recognition of the technical difficulties and costs of complying, the potential for error, and the stigmatizing effect of strict legal enforcement. Furthermore, the processes giving rise to pollution sometimes produce erratic discharges in which effluents may become heavily polluting, and river water is also inherently variable in quality. The result is that a degree of leeway is normally granted to dischargers, and a certain amount of pollution allowed to occur with impunity. Such leeway is the means by which the enforcement agency adapts to uncertainty. 'The present situation is that consent conditions are written as though they were absolute,' a senior official said:

> 'The time it goes over—once in a lifetime—and you're a criminal. So the practice has been to regard them rather unofficially as having been satisfied if there is compliance on about ... three occasions out of four, or four out of five, or two out of three—practices are variable from one authority to another. We've adopted a three out of four.'

Non-compliance with standards is thus organizationally sanctioned.[4] Though the consent may not be strictly enforced, however, it is significant as a benchmark for both dischargers and enforcement authorities. It remains the criterion of pollution, despite its irregular enforcement; as such it creates a zone of officially tolerated 'pollution' which will vary as the standard and the setting of the discharge vary. And it remains the legally enforceable standard of pollution if customary enforcement practices are suspended.

In a substantial proportion of cases dischargers normally display little or no objection to the standards imposed by the agency, as a result of preparatory work in negotiations conducted by the field officer together with, in more important cases, his area supervisor. But where standards appear to a discharger to impose excessive demands he may exercise a right of appeal to the Secretary of State. Although hardly ever employed, this right of appeal in effect confers bargaining power upon the discharger who seeks to counter administrative extravagance, since the agencies attempt to avoid appeal at almost all costs:

> 'It's certainly true that if people suggest they will go to the appeal procedure then we will do almost everything we can to avoid that, which means in some cases we're letting people get away with something that you wouldn't let others get away with, simply because they're being difficult—well, not being difficult, they would be just exercising their rights.'

The desire to avoid appeal may partly be a reflection of the fact that standards have been selected by resort to the convention of established—but largely unexamined—practice which may be difficult to defend in a formal arena. Thus 'you've got to be very sure of your ground, because you've got to get up, not in a court, but in an appeal situation, and give evidence'. There is also a good organizational reason why agencies fear appeals: appeals introduce uncertainty (cf. Kaufman, 1960:154-5) and threaten their control over the design of consents. An appeal may establish a formal precedent which an agency might find undesirable, since standard setting is then taken out of its hands by people regarded as possessed of less expertise and possibly differing interests. A principle informing negotiations about consents is to avoid the risk of creating a precedent on appeal injurious to the agency's interests, a principle of wide applicability, for to appeal and win—or even to appeal and lose—publicizes the possibility of remedy which itself is a threat to the agency's whip hand.[5] Appeal gives status and recognition to protest.

Ironically, recourse to precedent is a useful tactic for the field man to employ when negotiating about standards. Reference can be made to standards usually set by the agency 'in cases of this kind', or attention drawn to practice in past cases. But the officer must maintain a certain flexibility and be able to deviate from usual practice where necessary, especially to avoid the risk that an administrative convention may be established which can subsequently be used by others with conflicting interests as a resource in their own negotiations. Obviously when the agency claims allegiance to an established precedent where its formal interests in pollution control or its organizational interests are served, it lends further support to the principle of equity. Indeed, the principle serves administrative interests well, for if standard conditions designed to be equitable as between apparently similar dischargers are routinely applied, the discharger who wishes to appeal against the standards imposed in his case is probably in a weak position if the agency is able to show the Minister that the standards were those typically in force. But in practical terms, the organizational imperative to avoid appeal means modifying demands in the course of negotiations if there is any suggestion that the agency might have to defend itself in an appeal: 'we try to see', said a senior officer, 'there are no grounds for reasonable objection.'

II. Procedure

Standard setting is not a large part of water authority activity because most dischargers have long since negotiated their consents, and a lack of industrial development has depressed the number of new consents applied for. More important in recent practice has been the review of consents already granted, a matter which has revived interest among agency staff in some of the major issues surrounding standard setting.

Consents tend not to vary greatly even between apparently very different effluents. Limits are normally imposed on organic pollution and solids per unit of volume and often on ammonia, while industrial discharges frequently have an acidity or alkilinity (pH) parameter. Certain kinds of manufacturing processes attract specific parameters such as various metals, cyanide, phenols, or temperature.[6] In most cases the choice of parameters and limits is made by the officer in the field in response to an application from the would-be discharger,[7] a choice which is routinely ratified by senior staff.[8] 'I just look at them,' said a senior official. 'I don't calculate them all.'

Once the field man and his area supervisor have settled upon the standards to be imposed, the consent is approved by the head of the agency and ratified by an advisory committee. Staff senior to the area supervisor play an active part in shaping the standard to be set only rarely, in cases involving what a supervisor described as 'very contentious issues: possibly major discharges from the Authority's own works or from industrial activity where there has been a history of wrangling and debate about standards from that particular section of the industry'. Those very large industries which produce a substantial volume of effluent will also have their consents individually negotiated in recognition of the greater potential impact on the watercourse. Otherwise, the usual approach is to apply the so-called 'Royal Commission' standards[9] of twenty parts per million (p.p.m.) biochemical oxygen demand (BOD)[10] and thirty p.p.m. suspended solids, with the addition of other parameters as appropriate. It is done intuitively. 'If we're talking about BOD or suspended solids, we have a feel for that ... you make a "gut reaction" decision.' The choice relies heavily upon administratively established convention—'normal ways'—as a device for help in determining any case at hand. 'I think the first thing I tend to do', said an area supervisor, 'is ... to look at the normal working standards normally applied and see whether they would be adequate to protect the river; and you then take into account the flow, the volume of the discharge or the rate of flow of the discharge and the volume.' In the majority of cases the process is unexceptional:

'It starts at [field] officer level. The ... officer will probably say "Royal Commission", or whatever it should be,or something 'long these lines to some industrialist. The area man'll look at it and say "You can knock them down". "There might be oil from this site, we're putting oil on as well." Or," I think that one's a bit too strict." Or he would perhaps modify it. And that's probably it.'

There are also tacit constraints upon standard setting. Agency staff sometimes question themselves about a discharger's economic position and his capacity to comply with the parameters and limits which they would ideally like to impose. But predictive judgments about ability to comply are founded not upon some dispassionate analysis of economic facts and figures so much as upon characterizations of a discharger's know-how and willingness to comply, derived from his occupation, size, experience, and reputation. 'I think to myself "Well, has this company got the expertise to treat to this particular standard?",' said an area officer:

' "Are they likely to be able to attain it and maintain it? Are they, will they have the expertise? Can they be trusted?" Y'know, I mean there are companies and companies. You expect that if you set a 10:10 standard[11] on a [large well-known company] for a, say a pickle liquor, or something, if you did that they would comply because they're big enough and they've got the expertise, and so on. If a company called Joe Bloggs and Company that were from a back street ... came with a similar proposal and they obviously hadn't got two ha'pennies to rub together and certainly weren't gonna employ a chemist or anything like that, then ... you would view that as a completely different proposition. You would probably say "No. We won't issue you with a consent at all." '

These remarks emphasize the central judgment of reputability. The large, well-known company possesses 'expertise' and can be 'trusted', while the reference to employment of a chemist reveals that the small company lacking the attributes of wealth and competence is less able to comply, and implicitly untrustworthy. In the absence of constant monitoring, trust is of crucial importance to an enforcement agency's detection and compliance processes.[12]

Difficulties sometimes arise when the agency redefines pollution control policy for a stretch of water and wishes to tighten existing standards. Some of the issues typically raised during standard-setting negotiations surfaced in the following case from my field notes:

McDonald's[13] is typical of an old well-established firm which has been absorbed into a larger manufacturing group and modernized. The company makes a wide range of bathroom ware and discharges its effluent into a fishless urban stream. The authority's predecessor river board granted a consent some years ago with a number of parameters: 20 BOD, 100 p.p.m.suspended solids (relaxed from the usual 30), and various metals, including zinc, consented to the standard 0.5 p.p.m. Laboratory analysis for the presence of various metals was not routinely carried out, however, until a couple of years ago, when it was suddenly discovered that McDonald's effluent contained 256 p.p.m. of zinc, a by-product of the glazing process. [Water heavily polluted with zinc paradoxically looks particularly clear.]

The agency demanded that McDonald's take remedial measures. Since then the firm has managed to bring the amount of zinc in the effluent down to about 20 p.p.m., but the agency has made it clear that it has to go below that. Dennis Blake, the area supervisor, wants to bring the zinc content much lower. He and his field officer, Brian King, are working to a timetable of compliance. After a slow start which prompted the field man to threaten to take a 'stat'[14] Blake now considers the firm to be abreast of the schedule he and King have set. Blake is satisfied that the threat of a formal sample was enough to put the wind up the company. McDonald's meanwhile went over Blake's head to complain to the director of the agency about the demands being made of them. The director, however, supported his staff; in fact, Blake said, he 'put the fear of God in them. You notice how they still turn white at the sound of his name.' [Later in the day during negotiations Blake will several times refer conspicuously to the director by name.]

As we drive to McDonald's, Blake explains that the company can probably only come down to about one p.p.m. zinc, and only then if it spends a lot of money. But it is technically possible. He predicts that McDonald's will ask for a consent of ten p.p.m. He says he will respond by asking for three, as a negotiating position, but is prepared to go to five. He has written a letter to his director of pollution control, suggesting two or three p.p.m., which has been returned with the annotation 'proceed as indicated'. Blake is adamant that he will not allow McDonald's to go above five p.p.m. To do so will effectively set a precedent for other dischargers who will soon discover that McDonald's is operating at an advantage. McDonald's belongs to a federation of companies in the same business and the area man takes it for granted that the firm's competitors will soon hear about the relaxed consent and apply to the agency for similar leniency. And if the agency subsequently has to treat other similar dischargers equally, Blake is concerned about a cumulative deterioration in water quality.

Blake has brought along a colleague who is an expert in the cleaning of various effluents to act as a consultant in the negotiations. He discusses with Blake the various treatment methods for zinc which the firm might try, raising the possibility of a sand filter. Blake is asked what a sand filter would cost. He says he doesn't know, then volunteers two thousand to five thousand pounds. [This is a fairly typical response of pollution control staff who prefer to err on the side of caution, in the form of wide ranges, when asked to estimate costs of pollution control remedies. Later that day, when involved in negotiation with the firm, the area man will adopt the lower figure for purposes of argument.] When I ask if the firm can afford it, he says it can, 'because of the size of the company', since it is part of a larger group. 'It's a lot of money to me personally. But it's nothing to a firm like McDonald's.'

We are met by one of the directors of the firm who takes us to a showroom for coffee. Conversation is soon transformed into preliminary negotiation. The director asks for ten. Blake says how pleased he is with the firm's improved performance, acknowledging the considerable efforts and progress it has made. But the legal requirement is 0.5. [A consent, once formulated, can be portrayed for purposes of negotiation as a 'legal requirement'.] He is, however, prepared to recognize that it is unreasonable to expect McDonald's to make one p.p.m. But he cannot go to ten. Blake talks of fairness to other manufacturers and observes that a discharge of ten will give McDonald's 'an unfair advantage'. [He will say later, 'If I'd relaxed the standard there I'd have had umpteen other people knocking on the door wanting to do precisely the same thing to save on trade effluent charges.'][15] The director nods in agreement and does not seek to argue the point. He has arranged for the agency staff to have lunch with the Managing Director and two experts in pollution control from the parent company who have travelled a considerable distance to attend. In the meantime, we are taken to inspect the treatment plant, where King takes a sample of the effluent. The engineer in charge reports that the firm has made great efforts recently to get some new equipment, but there has been a delay in delivery. [This is a story which pollution control staff hear regularly. In this case it seems not to be treated as a delaying tactic.]

We are then taken on a tour of the factory while waiting for lunch. Blake and King begin to complain to each other that they are now wasting their time, since matters are settled so far as they are concerned. After the tour we are ushered into the board-room where a magnificent buffet lunch is laid out. There is a splendid display of drinks. We are invited to help ourselves. Other directors join us, making a party of sixteen. There is talk over drinks until the Managing Director arrives. The pollution control men are not impressed with the lavish reception, King observing as he eats, 'They're trying to buy us off now.' Blake agrees: 'They're trying to soften us up. You enjoy it, but at the end of the day you have to ignore it.' He goes on to mention the occasional embarrassing moments which occur when firms try to pass backhanders.[16]

During lunch the Managing Director begins gently sounding Blake out, but the real business of the day does not begin until the table is cleared. The Managing Director makes a statement in which he adopts the position already put forward by his colleague, who meanwhile supports the various assertions with evidence. The statement culminates with a request for a consent to discharge ten p.p.m. of zinc. Blake repeats his earlier position, summoning the notion of equity between dischargers as a bargaining tactic to support his case, which seems to win the tacit approval of some of the directors present. For illustrative purposes, he uses the example of another similar manufacturer (whose identity, he reports, he must keep secret) who is able to conform to a one p.p.m zinc requirement. After each side has declared its position, the negotiations are diverted to more technical questions about further remedies which might be tried, and then adjourned.

[The company, after persistent pressure from the water authority, was able, with improved treatment methods, to comply several months later with a standard of one p.p.m.]

III. Enforceability

Though the agencies do not regard standards as strict limits, their enforceability is important. From the field staff's point of view it is essential that the parameters in any consent should be clear and unambiguous. They should at least appear to be attainable, for standards which can be met, if at all, only after massive expenditure might tempt an unwilling discharger to question them by appeal. Unattainable standards, even though not appealed by a discharger, may still lead to practical problems of enforcement, since a continued failure to comply because of inefficient treatment plant or other 'good' reason which enforcement agents tolerate may suggest condonation by the agency of the continual breach of consent. Indeed in some cases in which there is a persistent failure to comply and no apparent harm to the watercourse, agency staff tend to assume that the standards may be too tight, as a senior official suggested:

'somebody who's not complying with consent conditions and yet is not having much effect on the river, it probably means the consent condition is not right to start with. ... This is I think one of the reasons that we can tolerate some non-compliance because some of the consent conditions are a bit too strict.'

In areas of regulatory control like pollution, where the law regards conformity as essentially a matter of scientific measurement, deviance is conspicuous. While clear, uncomplicated standards are useful for an enforcement agent negotiating to secure compliance, economy in the design of standards is also important. To be parsimonious in the use of parameters assists in the display of deviance; in a prosecution it aids a portrayal of guilt, as an agency head explained:

'I always try and have as few conditions as possible in a consent. ... In a court it would be so easy if I was on the other side to get up and say that in respect of eight of the conditions "We complied and we only contravened the suspended solids condition and we're very sorry it was 30,000 p.p.m. but you're not gonna cane us for that are you, Mr Magistrate? We've got eight of them right." ... So the simpler the conditions the better, the fewer conditions the better.'

Yet the use of few parameters, if they are well chosen, can still be an efficient means of pollution control. To have one or two strict parameters often means that other pollutants, which can remain unspecified in the consent document, are brought under control. Or parameters which provoke the discharger's concern can be framed leniently, control being assured by the insertion of a much stricter parameter of more general application.

The issue of enforceability was given a poignant twist following passage of the Control of Pollution Act 1974 which made an inroad into the virtual monopoly of pollution control enjoyed by the water authorities. The Act (much of which is not yet in force) requires publication of sample results and permits members of the public to bring legal action against a discharger who fails to comply with consent.[17] The agencies will have to have available for public scrutiny a register giving details of consents granted and results of effluent samples.[18] A source of particular vulnerability for the agencies is that in its reorganization of the water industry the 1973 Water Act gives them responsibility for the management of the great majority of sewage treatment works, which in many areas are themselves significant—often principal—sources of pollution. The water authorities thus not only set and enforce pollution standards, they are major polluters themselves. Yet the pollution control sections of the agencies are ultimately impotent to secure the compliance of their colleagues in sewage treatment, lacking recourse to the legal sanction (even if it were organizationally possible to prosecute fellow-workers). Many sewage treatment works have been consented in the ritual process to the usual 20 BOD and 30 suspended solids standards, which they have regularly failed to meet. While the agencies maintained almost total control over the enforcement of standards, a control which could be preserved in the absence of publicity, this posed no problem: 'A lot of the standards were set by River Authorities; and the works weren't capable of achieving [them] anyway,' a senior official said. 'But the former River Authorities knew this and shut an eye to it.'

After passage of the 1974 Act, however, the authorities had to contemplate the prospect that they themselves might be prosecuted for the polluting effluents from their sewage works. The new-found public accountability has produced a quite dramatic protective response from the agencies, illustrating the familiar point that a previously unreachable ideal may be achieved if reduced within the compass of the practically attainable. The agencies have simply embarked on a large-scale programme to revise consents both of their sewage works and of other discharges. Where a discharger would consistently do better than his existing consent, standards have sometimes been tightened. But where a discharger has regularly been outside consent, standards have been relaxed to accord with the existing performance in 95 per cent of samples taken. Thus while standards were relaxed, this was accompanied by less tolerance of their breach, compliance being expected in 95 per cent, rather than 75 per cent, of samples: 'exposing ourselves five per cent of the time', as a senior man put it, thinking of his agency's sewage works. The review was described by an agency director as

> 'an attempt to span out 95 per cent of the quality of effluent at present discharged from the [sewage] works, taking out the fliers and the obviously bad discharges. ... This means in practice that one out of every twenty samples will fail to comply with the standard and the Authority will be at risk from somebody, some member of the public, prosecuting.'

This protective behaviour has occurred in reaction to the threat which publicity poses to the discretion with which the agencies would treat deviants—themselves and others. 'We've learned in the last go it's the public accountability thing,' said a senior offical.

> 'The legal thing can go out the window. For the first time we've had to get our heads down and look at what we're doing. ... We bend over backwards to do the thing [properly] ... but what happens in practice—do we mean it? And in many cases we realize we don't. We find a factory who's not met its standards eight times in a year, and what do we do about it? Nothing! ... So one reason [for reviewing consents] is it doesn't make any difference to the river system; another is that it means we're not prepared to follow the thing through on a legal basis. ... It's going to be stupid when that booklet's published [giving details of sample results]—we'll get tons and tons of phone calls. ... What we're now trying to do is say the river will stand a temperature of 30°C—if it'll stand it, don't let's bugger about. It takes some sort of public document for people to get you out of a rut.'

Adaptive action by recasting the rules was necessary to preserve administrative control and avoid the public embarrassment of the authorities. Since financial, technical, and geographical constraints did not permit efforts to be made to bring their sewage works into compliance with the existing standards (even if time had been available), the only other means of demonstrably maintaining compliance was for the agencies to change the standards to fit the existing discharges.

For field staff, however, a review which led to the relaxation of a large number of consents appeared to be in direct contradiction to their efforts ostensibly to bring all dischargers into compliance—'trying to fiddle the books for when the Act comes into force' as a northern authority officer put it. For many, the task of policing relaxed consents was a source of considerable professional embarrassment following their attempts to secure compliance with the former, stricter, standards. For the pragmatic field officer, the exercise was purely cosmetic, not an effort to overhaul consents to place more rational demands (in terms of what a river could accept by way of pollution) on the discharger. 'To my way of thinking,' said an officer of wide experience, 'this is just paper—it's only figures on paper, this is—to prevent the general public being able to prosecute the authority if a sewage works has the odd bad sample.'

IV. Postscript

Pollution control standards are flexible markers of rule-breaking. How strict or lenient enforcement agents conceive them to be may well affect their enforcement behaviour. For example, working conceptions of cause may depend upon a judgment as to whether the standards breached were generally attainable at tolerable cost. If they were, a violation may prompt suspicions of deliberate wrong-doing in the interest of financial saving.

Until recently, pollution control standards were treated as guides rather than boundaries. They were a species of organizational rhetoric (now menaced by public access to the monopoly of enforcement), embodying compromise between conflicting values and recognition of the vagaries of the environment to be controlled. Standards which cannot be attained by negotiation or legal enforcement may at least be achieved if made less demanding. The threat of the 1974 Act was the challenge posed to agency control over the enforcement of regulatory deviance by publicity and a strict liability law. By redrawing the individual boundaries of deviance the agencies have sought to retain control. Their response may be the occasion for added impetus to the ostensible shift from equity to efficacy as the guiding principle in the design of water pollution standards.

II. Pollution Control Work

Three chapters are devoted to the setting and tasks of pollution control work. Chapter 3 describes the staff, the nature of the job, and the skills and knowledge it requires, and adds further detail to the formal organization of pollution control. How the organization impinges on everyday work is the theme of Chapter 4, which analyses the strains that arise when bureaucratic strategies of control are imposed on a job which demands a high degree of personal discretion at field level. The descriptive analysis of pollution control work is carried through into Chapter 5. Addressed here are the socially organized ways in which dirty water comes to be noticed and defined as a pollution, thereby creating a case about which an enforcement agent has to take some action.

3. Field Staff

I. Officers

Most routine pollution control work is conducted by staff—almost exclusively men—in the field officer grade.[1] Recent expansion of pollution control activity following the creation of the water authorities in the Water Act 1973 has resulted in an influx of younger recruits whose backgrounds and conceptions of the job contrast to an extent with those of older, longer-established field staff. While it would be imprudent to press distinctions too far, since the practical demands of the work impose their own priorities, field officers themselves are aware that there are virtually two generations of staff, each with a characteristic view about pollution control work.[2]

The newer recruits who joined the water authorities after reorganization of the water industry are predominantly young men in their twenties. They have enjoyed a higher level of formal education than their older colleagues, some of them possessing degrees in chemistry or biology from universities or polytechnics. For some, working for the water authority was their first job; few have any substantial prior experience in industry ('working on the other side'). They are spurred by a marked sense of vocation: they are of the generation which grew up after the war when pollution came to be popularly identified as a serious problem needing urgent attention. And they are ambitious: while the job is regarded as important and worthwhile, it is for them, only a first step in a pollution control career, even though the present organizational structure of the agencies provokes anxieties about the limited prospects of promotion.

The younger field men, doubtless reflecting their education in the natural sciences, tend to conceive of pollution as a relatively unproblematic phenomenon, as scientifically determinable, hence 'obvious'. For them pollution control work is a scientific endeavour. Accordingly, the design of standards and the securing of compliance are matters for the dispassionate application of scientifically-derived principles; and the formal legal process is the primary instrument by which control is to be maintained. Their stance is typified by one young officer's remark, when faced with a recalcitrant discharger, that 'all cases should be looked at as potential prosecutions'. Heavy reliance on negotiating to solve problems will sometimes provoke the criticism from them that too easy-going an approach leads to inefficient enforcement.[3]

Most of the older officers have worked in pollution control for several years and many have had a number of years experience in industry, which, they are quick to point out, affords them an 'insider's' sympathetic understanding of the difficulties and costs which arise in complying with standards. In contrast with their younger counterparts, some of whom at least could hope realistically for the possibility of promotion, the older field men have little ambition, though they retain some sense of mission. They have been passed by, and know it. Some resent the emphasis increasingly being laid upon formal qualifications in the authorities' recruitment policies. With no high level of education they lay claim instead to 'experience'—an intimate familiarity with the way things have been done in the past—and 'common sense'[4]—a rich fund of practical knowledge about the job and its problems which inculcates a sense of the reasonable. In an occupation depending heavily on the officer's craft, the value of education and 'bookwork' is dismissed. A degree is no shortcut to the acquisition of the essential practical skills; experience is won by time and application:

> 'I know today now there are some good degrees and goodness knows what for the job, but it doesn't mean to say you can do the job. You might be able to learn it parrot-fashion out of a book or be able to put it on paper. That's something I couldn't do. I couldn't write my job down on paper. It's about eighty per cent experience and the other twenty common sense.'

Some field men have been in the job long enough to have seen the protection of the environment become a matter of considerable public concern. They, meanwhile, continue to work much as they have always worked. For them pollution control is less the application of scientifically-determined standards, more the art of managing personal relationships, 'a public relations exercise'. While newer recruits prefer to pin their faith on an 'objective', 'scientific' conception of pollution standards which should—in theory at least—be rigorously applied, the older men, with relatively few exceptions, prefer to 'be reasonable' and to 'understand the polluter's problems'. Rejecting the younger officers' professed greater commitment to prosecution as a ready sanction for deviance, the older staff claim they will negotiate whenever possible, contemplating the law as the ultimate response, resorted to with extreme reluctance.[5] The younger men are sometimes accused of being too officious, too ready to throw their weight around:

'They know they've got the law behind them. And you've got the odd youngster who will go there and throw the book at somebody. ... And this gets their [polluter's] back up and they don't want to know you then. But I find this with quite a few of the young ones. This is the first thing they want to do.'

An authoritarian stance is taken to be evidence of the younger staff's lack of experience of the harsh realities of industrial and agricultural life which has denied them the appreciation of the 'insider'. Their approach is not simply regarded by the older men as a reflection of youthful zeal and an extravagant sense of mission, but as a means of adapting to the difficulties, as young men, of presenting themselves as possessed of legal authority (an interesting practical recognition of the importance of interaction in securing compliance). After all, as an older man said, 'It could be that some of these farmers don't want to be told by someone who's about twenty-two years old.'

The older men have witnessed many changes in the formal organization of pollution control which have had implications for their work. They have had to come to terms with computer print-outs, data sheets, the use of electronic equipment, and biological sampling techniques. Most significant for them has been the need to adapt to the much greater degree of bureaucratic[6] organization which has followed the 1973 Act, the great landmark in their careers. The character of the field man's work was fundamentally altered in ways unknown to younger staff. Even so, a sense of vocation is noticeable among most field officers, even many of the older ones: pollution control is still conceived of and practised more as a calling than a job.[7] Indeed, it is the sense of mission, so far as the older officers are concerned, which produces resentment in the face of efforts to bureaucratize the job. Despite the older officers' complaints, the level of job satisfaction—when they can set aside the paperwork and get out into the field—is high (cf. Ross, 1970:38).[8] Each district is different, but each is tended with a considerable degree of commitment, field staff speaking fondly and possessively of 'my patch', 'my stream', 'my trout'. The nature of the task continues, despite bureaucratization, to provide for a significant measure of autonomy.

The view of pollution control as a practical art acquired by experience defines the particular qualities demanded of a 'good' field officer. Fieldwork, nevertheless, involves the routine application of a wide range of technical and scientific knowledge. An officer must know some chemistry and biology, and the local geology and geography. He must, in particular, be familiar with the science of water treatment and purification processes, and well acquainted with manufacturing and agricultural techniques in order to understand local industrial and farming problems. Knowledge of the law relating to control of water pollution, beyond a broad conception of the pollution offences, however, is regarded as unimportant and is claimed by only a small minority of field men, because the job is done 'by experience' and the application of rules-of-thumb—not 'by the book'. As long as they know what kinds of acts or events might be prosecutable, they take the view that the arcane business of the law is best left to senior staff. But the real job (rather than the ideal version articulated by younger men) is, as we shall see, much less about the application of science and technology to the measurement and control of pollution than the use of negotiating and bargaining skills to secure compliance. Ultimately water authority staff prize personal qualities as an officer's most important attributes.

A substantial amount of a field officer's time is spent in creating and preserving good relations with dischargers of all kinds, even in the absence of pollution problems. Officers are dependent upon their dischargers to do their job efficiently and must always think to the future when they may need the active co-operation of a polluter. The art of managing personal relationships, and the ability to talk independently to people with tact, diplomacy, and persuasiveness are great assets. Scientific knowledge counts for little in comparison, for it is the field man in the great majority of encounters who is the 'front man', the public face, of his agency (Goffman, 1959). It is he who represents and transmits the policies of the agency to a predominantly reluctant community:

> 'I think it's important that you should be presentable. And you must have a way of being able to get on with people. This is *very* important and I should imagine there's been more trouble, probably, with people representing the authority and being rather officious, and this has got no one anywhere. ... And I think you've also got to have people who do bear in mind that they're about one of the few parts of the authority that are in the forefront of the people and are basically representing the authority. So you must be careful of what you say. ... There's more of public relations in this job than probably—— that doesn't come down in print. ...‡ I think the better you are at public relations the better you can do your job. You might not be technically the best man on the job, but there's a whole back-up organization for this.' [His emphasis]

Negotiating skill is more than simply being able to 'get on with people', it demands the art of managing an encounter with a polluter.

‡ I use a long dash (——) to signify a pause in the talk. Points (...) represent deletions.

It is essential 'to appreciate other people's points of view but to be able to swing them round to yours if necessary', a social skill which must be practised in an extremely wide range of relationships. To be able to do this creates a context in which effective bargaining can be conducted. Dischargers are regarded as potentially difficult people. Each is confronted with different problems, some more intractable than others; each is in a different economic position. And the commitment of each discharger to pollution control will vary widely, depending not least upon the organizational status and occupation of the field man's contact. All of this must be accomplished while the officer continues to do what he sees ultimately as a policing job. If he is less than successful he may find that he has to work with an unco-operative discharger. Good rapport cannot be assumed; it has to be cultivated and regularly tended. The field man has to talk to people from a variety of backgrounds, in a variety of jobs. 'They've obviously got to be able to speak to people at all levels,' said a supervisor of his field men.

'They've got to be able to talk to workmen, and get things done during pollutions—find things out. You find an awful lot of things talking to Fred who works the filter press, y'know. And at the same time you've gotta be able to talk to the managing director and have lunch with him and talk about what was on television and what the state of the country is and the economic situation and everything, you know. So you've got to be able to talk to people. You've not gotta be timid, y'know, because obviously people will try and browbeat you and intimidate you if they think they can.'

Part of the field officer's art is to display an understanding of each polluter's problems, and to resist the temptation to treat as routine demands which may prove to be extremely onerous, particularly for one in an economically vulnerable position. It is not simply a question of the officer displaying an ability to be sympathetic in his demands, he must treat as many cases as possible on an individual basis. He must always remember he is a 'repeat player' (Galanter, 1974). As one supervisor said,

'You ought to be tactful and you have got to have initiative, because at times you have got to make decisions. They might only be minor, but to the person you're talking to it's a major decision. So you've got to take the initiative and be firm where you need to be with a company or with a farmer ... and consider his needs, if they be economical, impracticable. ... I think it also helps if you've ... got industrial experience so that you realize the problems of the industrialists.'

Another aspect of the craft to be mastered is the ability to conceal the conflicting demands of being a conciliator bargaining to secure compliance, while ultimately doing a policing job. The officer must indeed be prepared to sustain his attempts at persuasiveness over a long period of time, a matter which can pose problems, for he must be able to sense when a less conciliatory performance is warranted and a show of force, however discreet, would be appropriate. On the other hand, a display of due respect which denies the possibility of being interpreted as deference may be necessary when dealing with senior officials in a large and powerful organization.

And if a conciliatory approach fails to work, the job must still be done. Qualities of persistence and determination when confronted by the evasive or blatantly unco-operative now assume considerable significance. These virtues are especially important given the solitary nature of the job, as an agency head emphasized:

> 'The man in the field must be prepared to work on his own and work conscientiously. He's gotta stick at it, y'know, because if ... the discharger ... [says] something unkind ... you've gotta go back the next week, the next week, and the next week. So I suppose you've got to have a little bit of a thick skin. ...'

Finally, the solitude of the job also emphasizes a need for a high degree of personal integrity in each field officer. 'He's gotta be honest', said a supervisor.

> 'That's vital because he's in a position of great trust. He's operating his own hours. We trust him to work. We trust him to do the mileage he says he does. He's going into industrial premises: he's great possibilities for fiddling and taking hand-outs, pinching industrial secrets and passing them on. It's limitless. So honesty I think is absolutely vital. You've gotta be able to trust the man.'

Agencies are not, however, without various strategies of control over field staff (ch. 4, s. ii).

II. Doing the Job

Pollution control work is varied and unpredictable. Field staff operate in a range of environments each of which create distinctive kinds of task. In an industrialized urban area the job, especially in the northern authority, is essentially one of cleaning up heavily polluted rivers, but in rural areas, where there are fewer sources of pollution, field officers are more concerned with preserving high standards of water quality. The degree to which pollution is defined and tolerated will vary markedly, depending on the character of the officer's territory. However, although the problems posed by a densely-populated industrial area will differ in degree and kind from those in a large rural catchment with a scattered population and little industry besides farming, the fundamental approach is the same.

The job is essentially a form of policing, though in keeping with a compliance system of enforcement, tasks such as surveillance and prevention are emphasized. The formal processes of the law, indeed, do not loom large in the field officer's routine activities. It is recognized that the everyday work of monitoring and negotiating is itself an expression of the agency's legal mandate, and the legal rules about pollution may occasionally be invoked as part of the process of securing compliance. But it is rare to initiate the formal process. Yet the image of policing persists:[9]

'I'd say in a way that basically it's a form of a policing job. You are trying to police without wearing helmets. You are policing rivers is about the simplest way of explaining quality control. ... You are seeing where pollution is and you are policing. And the point is that wherever you go you tend to keep your eye open and not wait for something to crop up; you do tend to say "'Ello, 'ello, 'ello, What's this? Where's it come from? What does it mean?" ... If you're describing it, say, to laymen ... you've got to make the point that you are controlling. ... You really must put it over that it is your job to know every wretched pipe anywhere on this river system ... that it's your job to know what is going on, who is doing what.'

These remarks emphasize the role of surveillance in pollution control, for enforcement is conceived of as a matter of prevention as much as of responding to rule-breaking. In a typical week, the field man will spend much of his time in routine inspection and monitoring by sampling river water, industrial and farm discharges and occasionally groundwater supplies.[10] He will also regularly look over pollution control installations in his district to ensure their satisfactory operation and maintenance. General preventive work is equally important. Staff must inspect sites for which planning applications have been submitted in order to assess the implications for water pollution of any building or other development. The licensing of tip sites for the disposal of solid waste requires similar inspections. Trivia must also be attended to: drains unblocked, obstructed pipes cleared, leaks looked after.

Frequently the officer will engage in specific preventive work: issuing warnings to dischargers about risks being posed; offering advice about essential remedial measures; supplying information about grants which may be available; and suggesting consultants.[11] This is often a time-consuming but indispensable part of the field man's job since the work demands that he both preserve existing relationships with key people in farms and factories in his district and seek out and cultivate new contacts.

The enforcement relationship is symbiotic (Manning, 1977:136-7): officer and polluter depend upon each other for information and assistance (ch. 6, s. iv). Being on good terms with dischargers is essential for the field man to do his job efficiently: it allows him access to property whenever necessary, enabling him to carry out inspections, and to raise matters which might otherwise be sensitive. A good working relationship also enhances an officer's intelligence system, more effectively bringing to his attention knowledge of pollutions for which the discharger or his neighbours may be responsible.

In the first few years in a district the officer learns about his territory as thoroughly as possible. It is his reponsibility—his 'patch'—and, like a policeman (Rubinstein, 1973; Van Maanen, 1974), he must know it intimately. He will be expected to live in it, or at least close by.

The field man's mental map of his territory is one structured by rivers, factories, and farms. He must know the origins, courses, and confluences of all rivers, streams, and ditches, the rates of flow in dry and wet weather, and the uses made of the water. Topography adds another dimension to the subtle geography of territory, for the field man must also be aware of the contours of his patch, its slopes and catchment areas. He needs to be acquainted with the exact location of every industrial discharger, their manufacturing processes and wastes, and, in particular, the nature, volume, and flow of effluent. He must also know where farm effluents discharge and—most important of all in view of the volume of pollution they cause—where the sewage treatment works outfalls are.

A few polluters with ineffective treatment processes discharging effluents into otherwise clean rivers can make a patch 'difficult'. Hidden watercourses also pose problems; thus if he has an urban patch, the officer must know where the foul sewers and any culverted watercourses run (a formidable task in some cities), as well as the nature and source of any discharge into the watercourse.

Knowing the territory means acquiring a sense of the distribution of pollution by identifying the sources of trouble, potential and actual. This is another part of the catalogue of essential knowledge which officers, like the police, employ to refine their intelligence systems, assisting control by efficient discovery and detection. 'You should learn your river and your places like the back of your hand,' a senior officer said.

'I mean I used to. I could have at one stage recited to you every discharge from [the estuary upstream for fifty miles] on the north bank of the river, in order. Because the only way you do your job is to get on your honkers and walk round the pipelines and know where everything is and which are polluters, and which ones go up and down. And you've got to get there at night and at weekends to find out what really goes on, and whether they are meaningful in terms of pollution.'

Field staff regard their work in intensely practical terms. Pollution control is about action in the field and getting results. Anything else is secondary. For this reason office duties are disliked and paperwork despised as 'dirty work' (Hughes, 1971:343 ff.) which diverts time and energy from 'real' pollution control work,[12] making one accountable for matters which have little to do with law enforcement. But such work must be done, and field men put in several hours each week in their area offices, writing reports, keeping records up to date, drafting letters, answering the telephone and communicating with colleagues in the field and staff at headquarters. Every few days each officer can expect to take his turn on the office duty rota which will keep him away from the field for a complete day. Every few weeks he will have to spend a weekend at home near a telephone on call to handle any emergencies which crop up in the area.

The following extract from my field notes—warts and all—recording a day's work described by the field officer concerned as 'typical' of his district may help convey a picture of a pollution control officer's job in an urban industrialized area (rural officers are usually under less pressure of work). Many of the points touched on will be elaborated in subsequent analysis. The notes were taken early on in the research when I was chiefly concerned with documenting pollution control work, thus they should suggest something of the nature of the job—routine tasks as well as special problems—which an officer can encounter on a summer's day. The notes are reproduced in almost the exact form in which they were scribbled (any editing is to preserve anonymity):[13]

I arrive in the area office on Tuesday afternoon, having spoken that morning with the area supervisor at headquarters. I meet Kevin Griffith and Don Barry, both of whom view me rather suspiciously. They ask me about the research, its aims, methods, etc. I make arrangements to meet Griffith in town the following morning.

G. first samples the City's major sewage works effluent.[14] In the hot weather the works has been having some trouble in staying in reasonable control of its effluent. The authority is keeping a close watch on it with samples every day. Derek Andrews should be sampling but G. stands in during Andrews' holidays. One sample is taken of the effluent, with both downstream and upstream samples of the river.

G. visits a couple of pollutions which have been giving trouble in the last week or two. The first is in a park in town. It is a small ditch which emerges from under gardens backing onto rows of houses. There is a small amount of water in it, but hardly any flow. 'You can see what the problem is', G. says. It is very clear. Lumps of excrement and sodden pieces of toilet paper lie in the water in a state which suggests they have not been broken down at all. G. doesn't know where it's coming from, but waves a hand in the direction of the houses, suggesting a pipe may have been wrongly connected, or a sewer has broken. An example of the detective work which pollution control involves. He is unable to do anything and we move on.

The next problem is an ice cream pollution in another small ditch. Again there is low flow, although there is rather more water. There is a foul smell coming off the water, which, although cloudy, does not appear too dirty. The pollution had occurred a week earlier. G. had traced it to an ice cream works employing about six men. A complaint had been received—the works hadn't reported a pollution. When confronted with the pollution, G. says the manager claimed it was an accident, that a vat had tipped over accidently and the ice cream had disappeared down a drain. (Ice cream, like milk, is extremely polluting.) The man was extremely hostile and aggressive. G. decided not to take a stat[15] for this reason, though there was every reason to, as he acknowledged. G. says he thinks the man is lying, that the vat went over not by accident but because the ice cream was not up to standard and pouring it away was the most convenient way of getting rid of it. G. agrees that the offender has probably escaped prosecution because of his violent and aggressive demeanour which made G. think first of his personal safety. G. takes a sample in the ditch, which shows clear signs of sewage fungus, with its tell-tale grey strands waving in the water.

We go to the city canal, reaching it through an aluminium works. The works is closed down, like many of the city factories, for the holidays. Some men are still present, however. G. and I trek round to the sampling point for the discharge to the canal for the cooling water, which is when we see the plant is closed. By the sampling point there is an iridescence on the water which regularly fans out, indicating a seepage of oil. An oil tank behind with a three foot bund wall[16] all around in which more than a foot of mobile oil stands. G. assumes that somehow the oil is escaping into the canal. He decides he must do something about it and goes off to find the works engineer. A youngish man, the engineer seems a little concerned at the news. We go back to the site. Discussions about how best to check the flow. G. wants the embankment dug up so that the pipe can be examined. The engineer says he wants to drain the oil out of the catchment tank as there's no point in doing anything as costly as G.'s suggestion to begin with. G. accepts his suggestion and leaves, observing that he will be back fairly soon to see what sort of progress has been made.

We go off to another site, reached by a quarter mile trek along the main railway line. A large depression in the ground with a watercourse in which a certain amount of sluggish water stands. There is a heavy oil scum on the surface at one end, which has collected naturally. G. doesn't do anything, the purpose of the visit being to see what conditions are like.

We visit the river near the city centre. A purely urban river, rising in the south of the city ... flowing through the centre into the main river. We disappear through a locked door in the wall. I find myself on a small brick platform about twenty feet above a man-made, well-bricked channel which follows a straight course through the factories and warehouses. The water looks clear and has long green weeds waving in it. G agrees the water looks good and says it's a Class Two river,[17] except when it rains, when it gets all the run-off and also lots of foul water from the foul sewer overflow. Although the river is only one to two feet deep G. says he would not walk across because the current is too strong and bottom too slippery. He's fallen over in it before now. He doesn't take a sample but examines the water gushing through a drain from under the road, just where we've entered the river culvert. Looking up and downstream I can see lots of discharge points which would rapidly raise the level of the river in wet weather.

To a food manufacturer's in another part of the city and the adjacent brook. G. takes six samples in various parts of the works. Some are very inaccessible and he has to wade through thick mud to get at them. G. suspects that there may be sugar being discharged into the water. His liaison man at the manufacturer's is the works chemist. G. complains that he's not as helpful or responsive as some of the other people he has to deal with in other factories. He doesn't see or talk to anyone from the works this visit.

By now G. is no longer suspicious of me but turns out to be a lively, talkative, and intelligent man. ... He explains that the day so far has been pretty typical. As he does so he stops the car because he sees water in the road which is coming from a site on the left hand side. We go onto the site which is a ready-mix concrete works. The water has been used to wash lorries down. It collects in a pool near the gate, thence it trickles down the road. It is heavy with suspended solids. The canal is at the bottom of the road. The water seems to dry up on the road without getting too far. G. takes no action. Instead we walk along the canal bank 350 yards looking to see if there is a discharge point into the canal from the works. We cannot find one.

Back in the car G. repeats that the day has been typical except we haven't been called up on the radio and he's been called up twenty-nine times in the last month. At that moment the radio transmits G.'s call number, informing him that there is a pollution at a motor manufacturing works.

We arrive at the works and call in at their pollution control offices where they have a small full-time staff looking after pollution in the works. The works is enormous, employing many thousands. A youngish man deals with us, taking us to the shop floor where camshafts are made. He explains that some cutting oil has been noticed in a culverted drain under the shop floor. He responded by blocking off the pipe, which leads directly to the river, and by calling the authority's pollution control office. Entering this part of the works I notice two or three prominently displayed boards alerting employees to the phone numbers of officials who must be contacted when a pollution occurs.

We enter the factory itself. Lots of machines—very heavy, rather noisy. We inspect the drains via three manhole covers and establish that the oil is coming from a certain area in which there are a number of cam-cutting machines. The factory's pollution control officer is clearly demonstrating his concern to G. G. is also concerned, because the company had a similar pollution two weeks earlier which the local paper had headlined 'WHEN THE RIVER ... RAN WHITE'. The cutting oil emulfifies in water and looks the colour and consistency of milk. It needs a massive dilution before the colour goes. The local citizens of the area through whose grounds the river flows are pretty sensitive to pollution from the company. The local paper is also sensitive to the water authority ... it sees the authority as inefficient and costly. It printed a [bleached out?] picture of the river in its white state which over-emphasized the width of the river in a very misleading way.

The company man puts a green tracer dye in a couple of machines, but none appears in the drain. It seems to be a leaking machine, not a negligent pollution—as the previous one had been when workmen emptied oil down the wrong manhole. (This was part of the holiday cleaning-up. G. observed that although holidays mean a shut-down in industrial activity, they can lead to plenty of pollutions because of the cleaning that goes on in the factories. Many of the workers there are either ignorant or careless about where they tip their waste.)

That evening G. is phoned by the company and told the problem is solved and the leaking machine located. G. warns them he'll be round in the morning, to ensure, he says, that they keep making the effort to do a good cleaning job. G. is pleased with the reponse at the company—not only do they report as soon as possible when they've got a pollution, they let him know about the progress they make: 'I've got them so well trained now they'll be phoning half an hour before they have a discharge.'

III. Getting your Wellies Wet

Since the dominant conception of pollution control work is utterly pragmatic, doing the job is regarded as the best way of learning the job (cf. Van Maanen, 1973; Wilson, 1968). 'Common sense' and 'experience', the officer's greatest virtues, cannot be taught, it is believed, but have to be acquired by being thrust almost immediately into the daily round. So complex and unpredictable is the work that to articulate any formal course of instruction which would remotely address the variety and nature of the problems routinely encountered is thought hardly possible. Though one of the agencies has developed a number of guidelines about procedure to be followed in certain circumstances (see ch. 8, s. i), there are no standard manuals of field officer practice. 'You let them learn a lot of experience through themselves,' one of the older officers said. 'They gain more experience by tackling the job themselves.' The novice is dumped without ceremony into the field. 'It's the sort of thing you've got to start, go into, regardless of what qualifications you've got,' said a recent recruit. '... You get your wellies wet, so to speak, and you pick things up as you go along.'

The technical parts of the job—the sampling procedures, the use of chemicals and equipment—can be learned by going around with a more experienced officer for a week or so. Office skills—the ability to draft letters, to write reports, and in general to manage the bureaucratic demands—can be acquired in time from the area supervisor. These are, however, regarded as merely technical or clerical abilities. The pragmatism of field staff places much greater value on what a man can do in the field: his practical knowledge, his capacity to make decisions on the spot, his ability to talk to people. Fieldwork brings an indispensable first-hand contact with problems: 'You can't do our job in an office. I know one or two of them try and work it on the phone ... but I like to go out and have a look and know what's going on.'

These real arts of the field man's work can only be acquired over a long period, and they must be practised and refined by an officer on his own in the course of his work. The chances of learning directly from others are restricted. Opportunities to observe how the novice's area supervisor or other, more experienced, colleagues handle negotiations or use their discretion are limited and usually confined to the bigger pollutions or more intractable problems. If experience is gained from others it comes vicariously through informal contact in the area office. Otherwise the officer learns by his mistakes: '[you learn] from experience. I mean I've made one or two balls-ups before now. I try not to make the same mistake twice.' But this all takes time; for many, variety and the unexpected encourage the feeling that they are always learning.

The area office plays a prominent part in what a field officer is able to learn of the job. Whether a field officer works from his area office (in the southern authority) or from home (in the northern), contact with colleagues will normally be in the office, which he will visit every day, even during those occasional hectic days out in the field when he has to catch up on his sampling schedule. Most field men call at the office first thing in the morning to collect files, planning applications or other paperwork, and to pick up clean sample bottles or other equipment. Most staff return to the area office in mid-afternoon with samples to be sent off to the laboratory.[18] This is a time to catch up on urgent paperwork and swap stories with colleagues about the events of the day.

The office is the main forum for the exchange of experience, knowledge, and ideas. It is the site for regular meetings of all field staff with their area supervisor, at which problems can be aired, and official policy directives transmitted to the field. It is the place where the new recruit is most likely to learn about the 'typical' and 'difficult' kinds of cases. Knowledge about the particular problems exercising his colleagues has a practical importance if, as sometimes happens, an officer is called out in an emergency to a pollution in another man's patch.[19] It also has a broader value as a means of acquainting the newer recruits with 'normal' ways of dealing with problems. But what an officer will know of his colleagues' districts compared with his own will tend to be particularistic—of salient pollution problems—set in the context of an undetailed and generalized apprehension of the kind of patch a colleague looks after. In other respects a field officer will not claim familiarity with another district unless he happens to have worked it himself in the past.[20]

Perhaps the major influence in the development of a distinctive office culture, and for that reason an important contributor to a field officer's learning, is the area supervisor. In some circumstances the supervisor is in a position to visit his own views about policy and practice onto his junior colleagues, since he is the channel through which information and recommendations are transmitted up to senior staff and policy directives are funnelled down. Each area office is recognized to possess a special character which emanates from the personality and style of the area officer who presides over it. Conversations with both field and headquarters staff revealed a tendency to identify certain area offices and certain area men with a particular approach to negotiation and enforcement: 'the people at X office', 'the officers in Y area', or 'Eddie Osborne's people' describe in agency language an enforcement style which was in some way distinctive to the area.

A familiar feature of learning is a tempering of views.[21] Many of the younger recruits are regarded by older hands as extremely idealistic, a quality deemed more vice than virtue since too passionate a commitment to controlling pollution leads to aggressiveness in dealings with dischargers. Everywhere the novice looks he will see pollution. But sober reality will soon dampen the fervour of the most ardent. The typical new recruit regards himself, a supervising officer said, 'as a knight in shining white armour, which I think everybody starts off as. But after about six weeks it gets rubbed off a little bit when you find out the practicalities of the job rather than the, y'know, the theory.'

Experience, it is believed, leads to sound judgment, held to be a significant virtue in a field officer, for it is the field man who is the agency's gatekeeper. It is he who defines and selects those cases which demand action. The officer, an area supervisor explained, is

> the man who's actually called out in the middle of the night and stands in the pouring rain with his mac on directing operations. He's the one who gets the feel as to whether the company really are trying ... or not. And it all comes over in hard black and white in his report to me, and I've really only got what he says to go on, in that sort of case. If it's a one-off affair in some yard, the chances are I'll probably never ever go there, I'm really completely reliant ... on his reaction. ...'

To acquire soundness of judgment is essentially a matter of absorbing a sense of the norms operating in the agency which guide discretion at field level. Field staff, like other legal actors, make decisions in a context of what is 'normal' for 'cases of this kind' (Emerson, 1969; Sudnow, 1965). Learning how to categorize cases for the purpose of decision-making in pollution control may, however, be a lengthy process, since the solitude of the job and the absence of any formal training restrict the field officer's access to the agency's fund of knowledge about different types of case. The subtle acquisition of a sense of the agency norms—'the normal working standards'—is suggested in the following description of how an officer learned the rules about when to take 'a stat':[22]

> 'I think after you've worked about a year in the job you find that you start off by seeing pollutions. You think "I'll take a stat of that," and you find that when it's considered it's a fairly minor one, although it's strictly an offence and it doesn't comply ... it's way out of the normal working standards. It's illegal, in other words, but it's considered a fairly minor one, so we don't prosecute. And after you've taken several of these you think, "Well that's not a very bad one, I don't think I'll take a stat." So you tend to gradually come round to the view that unless it's causing very bad visual pollution, or you've had a run of complaints about it, then you tend not to take a stat.'

In effect, newly-recruited staff learn the normal working standards from their colleagues in the area office, and from their supervisor, who is in a position to exert some control over disparate individual views. The area office is the agent's only source of regular contact with colleagues, and a setting for the transmission of a shared system of norms of pollution control work. The office culture consists of a kind of folklore involving 'the selective preservation of the ideas that have proved worthwhile' (Shibutani, 1961:465), in part about 'normal' cases and in part about the most 'difficult' problems and the 'worst' offenders. In time, despite the independence of a far-flung inspectorate, newcomers are socialized into orthodox ways of doing the job, in which attitudes emerge and expectations develop in the cause of helping the work go smoothly. The folklore provides a version of past history on any case or class of case to which an officer confronted with a problem may turn for a solution, whether it be one of choosing the 'proper' negotiating stance, making the 'right' decision, or placing the 'correct' interpretation upon acts, events or characters.

IV. Reorganization

The reorganization of the water industry following passage in 1973 of the Water Act centralized and enlarged the structure of water pollution control in England and Wales. The result was the creation of large-scale public bureaucracies designed for the efficient regulation of water use. For enforcement agents the consequences of reorganization, with its greater degree of bureaucratic control, have reached down to the field and begun to change the character of the job.

Younger officers regard the change with a certain indifference. On the one hand they hear the criticisms of their older collegues and their nostalgic accounts of 'the old days'. On the other hand, they appreciate that the job has had a number of new tasks grafted onto it, adding to its scope and variety. Some of the changes, indeed, reflect their own personal preferences for a 'scientific' approach; science is an aid in their mission. 'I think it's developing into a profession because it's more technical,' said one young officer. 'The days of the old inspector walking the river banks are dying out rapidly, I think.'

Most of the more experienced officers, however, regard the consequences of reorganization with dismay. They resent the inroad being made into their autonomy which suggests impugned competence and a downgrading of status (cf. Prottas, 1979:155), and they dissent from the implication that pollution control can be done in an office, rather than on the river bank, talking to polluters and doing 'real pollution control work'. The 'feeling of freedom of action' (Blau, 1963:126) is what field staff most prize in their job, a feeling largely responsible for the high degree of job satisfaction they enjoy (cf. Crozier, 1964:204-5; March and Simon, 1958:95). The practice of their calling is now under threat. The growing demands to satisfy the appetite a greedy organization—the wider, more highly regulated sampling programme to be followed, the increasing numbers of reports and data analyses to be produced—are held to be unwarranted interference from headquarters, irrelevant impediments to the practical business of pollution control work. Many field officers are preoccupied with the increasing amount of time which must be spent in the office with paper—'keeping the organization at bay', as one of the more critical men put it. 'We get a lot of planning applications,' said an older inspector. 'We always had them, but not the numbers that we get today, nowhere near. And you get silly, what I call silly things. They want figures for computers and ... information to put on the computer, and this, that, and the other.'

According to the more experienced officers, the incessant demands for paper from headquarters divert them from more pressing tasks and display an innocence among senior staff about the nature of the job, a view the more firmly held in the knowledge that few senior men have ever worked in the field themselves (cf. Wilson, 1978:122 ff.). While regarded as able and industrious, senior staff are also thought to be remote, invisible, and above personal contact. Links between headquarters and field are becoming more and more depersonalized and bureaucratic: important decisions with major implications for field staff, like those to do with prosecution policy, are made 'up in fitted carpet land', where officers never tread. And when a decision comes down, it comes down unexplained to an officer who must continue his relationship with the polluter.

Centralized control by headquarters has forced field staff to yield some power over the routine disposal of cases. In the past the officer was free to a considerable extent to exercise his discretion as he pleased. In the interests of efficient enforcement his practice was to negotiate a solution to a problem privately, and at all costs to be patient.

Now—but only to a limited extent—the officer has lost some discretion in handling routine cases. The change is probably more apparent than real, but what seems to be an inexorable narrowing of the officer's range of effective discretion is demeaning to his sense of personal authority. Before reorganization, said an older officer,

> 'you were allowed to work on your own initiative. In the old days, if you had a district ... that district was yours. And my predecessor—even the [area supervisor]—wouldn't come in the district without letting you know where he was going. ... But now you never know who's running around in your district. ... You don't know what the higher-ups are doing. It was, if you like to put it that way, it was a status that it was your district.'

Some of the older men are also alarmed at what they see as a move away from conciliatory enforcement. 'Things were different in the old ... days,' said an officer in the southern authority. 'You had to have a cert, dead cert, iron case before any prosecution case came up. ... I don't remember a prosecution case in the old days. We always used to get it done through friendliness and chatting to people.' Another consequence of reorganization apparently detracting from job satisfaction has been a depersonalization of relationships within the organization. While field men still enjoy much the same degree of contact with their immediate superiors (the area supervisors), senior staff are now distant and unseen. In 'the old days' field officers (at least in the southern authority) enjoyed a high degree of personal contact with their superiors and were subjected to few demands from headquarters:

> 'In the old days the agency head knew every one of his staff. He knew their Christian names. ... You don't see that today. ... You've lost the personal touch with the higher management. ... Trouble with going back to the old days, the [agency] was more or less a family concern. You knew everybody. Everybody knew you. You go to [headquarters] now: if I go in there somebody says "Have you got your warrant card? Do you work here?" I been working here before some of 'em were born! It makes me seethe a bit when they say "You don't work here, do you?" They know very well I don't. I work a part of the river they've never seen.'[23]

Reorganization and the bureaucratization of the job have, in the view of some experienced field men, brought about a lessening of commitment, even of demoralization. One of the most vocal critics of reorganization, an officer of more than 20 years' standing, deplored what seemed to him to be an essential transformation in pollution control work. The younger staff 'want overtime or they want call-out money, or God knows what', he said,

> 'while a lot of the older staff, the older side of the staff, I think a lot of them are getting browned off with it. ... The loyalty to the job's been knocked out of you. ... I've noticed

this since reorganization. It never used to be like this before. ... It's this bureaucratic idea they're trying out that's getting the older staff browned off.'

The dismay among field staff about the change in the character of the job was neatly reflected by a printed statement posted on the wall in one area office, which nicely contrasted bureaucratic sentiments with blunter language about practical realities:

THE OBJECTIVE OF ALL DEDICATED EMPLOYEES SHOULD BE TO ANALYSE THOROUGHLY ALL SITUATIONS, ANTICIPATE ALL PROBLEMS PRIOR TO THEIR OCCURRANCE [*sic*], HAVE ANSWERS TO THESE PROBLEMS, AND MOVE SWIFTLY TO SOLVE THESE PROBLEMS WHEN CALLED UPON.

HOWEVER ...

WHEN YOU ARE UP TO YOUR ARSE IN ALLIGATORS, IT IS DIFFICULT TO REMIND YOURSELF THAT YOUR INITIAL OBJECTIVE WAS TO DRAIN THE SWAMP.

V. Postscript

Pollution control work is not regarded by its practitioners as a scientific enterprise in which a dispassionate discretion is informed primarily by technical concerns. Instead it is an art in which personal qualities are most important. The enforcement agent in a compliance system has a wide variety of roles to fulfil. He may sometimes be consultant and analyst; sometimes investigator and policeman; but he must also be negotiator and judge; inspector, educator, and public relations representative. This is because compliance is the end of enforcement and its effective attainment requires a constant display of helpfulness and reasonableness.

While the economic and legal implications of their activities are occasionally recognized, field men's discretion is shaped by the imperative to 'get the job done'. In doing the job they receive little formal guidance from legal rules or administrative policy. But this is not to suggest that their behaviour is unpatterned. Instead, ways of doing the job are broadly guided by some vaguely interpreted principles about what the organization's legal mandate is supposed to be and organized by the experience they acquire from their membership of an enforcement bureaucracy. Their work is practical; their knowledge concrete and particular.

4. The Job

I. Autonomy

Unpredictability is an enduring feature in the world of the police and other 'street-level bureaucrats' (Lipsky, 1976; Prottas, 1979). In such occupations autonomy is cherished as a means of adapting to an uncertain environment. Like the policeman (Bittner, 1970; Wilson, 1968), the pollution control officer enjoys considerable autonomy in his work as a matter of course. His occupation is solitary, inconspicuous, and largely free from direct supervision, leaving him at liberty to choose, within the constraints of his sampling programme, how he does his job.

As a reactive enforcement agent (see ch. 5, s. iv) the field man has no idea what will happen during the day—what complaints, what reports of accidents, leaks, or spillages will be telephoned into the area office. As a proactive enforcer a field man must always be prepared for the unexpected. In sampling and inspecting water quality he is his own agent to decide where and when he goes to pick up a sample. Establishing priorities in the sometimes confused jumble of things to be done is another expression of the officer's autonomy. In the absence of pressing problems he improvises his activities, working the case he wants to work.

The field man exerts substantial personal control over the handling and disposal of any case. The preservation of such control is important to the officer's own notion of efficiency. He typically determines the nature and level of the demands to be made of a polluter. He normally decides when an encounter with a discharger is appropriate and the kind of negotiation to conduct. He alone 'really knows' the problems. So much of the enforcement task is carried out in private negotiations between officer and polluter that the field man is in a position to be adaptive in the demands he makes, reflecting the exigencies of any particular case. Only in exceptional circumstances—a major effluent, a polluter who is proving difficult, or a discharger who is powerful or well known—is the officer's area supervisor likely to take a personal part in the proceedings.

Like other enforcement agents, the field man is gatekeeper to the apparatus of control. He is the person in direct contact with the real world of dirty rivers, the first stage decision-maker who normally controls his supervisor's access to 'problems' and 'troublesome' incidents: 'If we don't mention the fact to anyone higher-up, no-one ever knows. It's as simple as that.' The descriptions, definitions, accounts, and assessments received by the agency are his. Indeed, the higher a case proceeds in the organization, the more officials rely for their knowledge of 'the problem' upon a version constructed by field staff and substantially transmitted by the written word (cf. Ross, 1970:58-9). Senior staff depend very heavily therefore upon the man in the field, as one of them acknowledged:

> 'There's no doubt about it in my mind that pollution control is done by the [officer] on the ground. Everybody else in any organization, from the laboratory technician to the scientist to the director general, is there to support the man on the ground. If the man on the ground doesn't do his job correctly, the job probably falls down. Nobody else can do pollution control. We're all here to support him to some degree. He is the sort of——the roots in the ground, the bloke that really matters. ...'

The field officer's sense of autonomy derives from the fact that being allocated to districts on a geographical basis designed around river catchments gives him a personal territorial jurisdiction over which he presides.[1] One or two of his colleagues may know something of his patch if they have covered for him when he was unavailable, or if they have previously worked it before being transferred elsewhere. His area supervisor will also be acquainted with it, but will possess a detailed knowledge only of its conspicuous problems. In effect the field officer is the major repository of the agency's knowledge of each district. His sense of autonomy is also enhanced by his invisibility to others—he has no insignia of office, no uniform, or marked vehicle—and by the mobility essential for the efficient patrol of a district which will be geographically large (if rural) or complex (if a more compact urban location). The field man spends much of the day in his car. It is his own private vehicle, and, as such, can be made inconspicuous when he so chooses. It serves as a means of transport, a temporary office for writing notes and making records, a refuge in bad weather, a place to eat lunchtime sandwiches, and a store for the equipment: the sample bottles, sampling buckets, boxes, crates, chemicals, inflatable booms, ropes, leggings, boots and waders, and various other pieces of protective clothing.

It is out in the district that the real work is done. Here the field man is on his own. While his job lacks the other features of the policeman's work which lead to social isolation (cf. Skolnick, 1966:49-51), the pollution control officer is, like the policeman, a loner. Though he will frequently be engaged in discussion and negotiation with water-users and others, when it comes to making decisions and taking action the field officer is on his own. The job, accordingly, is one in which qualities of initiative and the ability to make decisions on the spot are prized—just as a policeman must be able to 'handle a situation' (Wilson, 1968:31). After all, pollution is mobile; the field man who hesitates risks losing his evidence.

Relative invisibility and freedom from direct supervision liberate field staff from some of the constraints found in other occupations. An officer who wants to take it easy, for example, or run personal errands can do so with virtual impunity.[2] Again, American experience suggests that field staff, like the police, could also be the target for gifts or bribes which polluters may make.[3] The field officer, after all, has the power to make a discharger spend a substantial sum of money. Though the topic is, of course, one particularly difficult to research, very little evidence emerged that attempts were made to co-opt field staff in this way. In one case a field officer inspecting a large garage which had been causing problems with an oil discharge was given an extensive tour of the premises and an offer of (unspecified) help with his car (this case, together with the lavish hospitality of the McDonald's case described in Chapter 2, were the only examples of an inducement, however vague, being made in my presence). Conversation with field staff would, very occasionally, yield instances of efforts which polluters had made to have colleagues accept small items, such as a joint of meat, drinks, or a free meal. The most striking example of an (unsuccessful) attempt at bribery was described by an officer who was offered a free choice of goods from a manufacturer's showroom if he would pour away a formal sample he had just taken (see s. iii, ch. 6).

None of this is surprising. Field staff, with few exceptions, possess a considerable sense of vocation and idealism, and display a commitment to their ultimate goal of the protection of the environment. Pollution control work also differs in important respects from routine police patrol work. The field man has to suffer less of the unpleasantness involved in policing. He is rarely out at night and is largely free to organize his activities so that in bad weather, for example, he can catch up on paperwork in the dry and warmth of the office. The work is rarely dull or nasty, like police patrol, and consequently field staff are not normally impelled to engage in the policeman's 'easing' practices or taking of perks to mitigate tedium and unpleasantness (Cain, 1973:36-7). Routine tasks can be easily interspersed with something more interesting.

The nature of the job, indeed, is such that tedium is shortlived. Its unpredictability, freedom, and diversity are its appeal. Even the humdrum tasks are varied enough to militate against a sense of monotony. One officer who had also worked as a trade effluent inspector remarked on 'the variety and the flexibility and freedom that this job gives you as opposed to trade effluent control. ... You're working ... within very tight guidelines in trade effluent, whereas a lot of this is your own initiative and there's that much more flexibility.' Controlling pollution in direct discharges to watercourses is very different, he continued, since

> 'trade effluent is very specifically geared towards discharges to foul sewer and the controlling and the charging of traders who are discharging into foul. Whereas this job, although it centres around the similar sort of thing, has become much more diverse particularly in the last three or four years with our involvement with deposit of poisonous waste and site licensing ...; planning applications are referred to a lot more now than when I first started. When I did start in '73 it was very much the same as trade effluent but the sewers were open, if you like. But since reorganization it's become much more diverse and we're expected to do a lot more different things.'

As one of the younger officers who came into the job around the time of the reorganization of the water industry, he enjoyed the additional tasks, in contrast with the more experienced men who feel their autonomy has been eroded. Much to their chagrin, those who look fondly back to earlier days when things were different, believe efforts by senior officials to regulate the activities of field staff more closely have created an unnecessarily ordered and bureaucratized job. Attempts to further the degree of contact between field and headquarters staff through more extensive reporting, and efforts to make their job more 'scientific' in their view merely restrict their occupational freedom. Thus, paradoxically, changes introduced ostensibly in the interests of rationalizing field officer activity are condemned as irrelevant and unnecessary attacks upon field level autonomy impairing the very efficiency the agency seeks to foster.

Recent organizational changes are, for the more experienced officers, evidence of serious misunderstanding at headquarters about real pollution control work: 'If you spend too much time in an office you can't do your public relations—you can't respond promptly to protect the good name of [the authority].' Some of the older staff have reacted by attempting to do the job much as they have always done it, which means, in effect, engaging in various minor acts of deviance from new bureaucratic demands in order to minimize the trouble caused. For example, to commit an account of an incident to paper is to endow it with a permanance and visibility to senior staff which may result in the creation of further—seemingly unnecessary—work. For the field man who deals with *real problems,* paper work is incidental, after the fact. In putting a case on paper the officer is also yielding control over its disposal:

FO Basically I do exactly the same now as I did then. But they made it—it's been made—more complicated.
KH In what way?
FO More paperwork. They want more reports. ... If you've got an oil spillage they want to know ... the make-up of the oil, what it is, and God knows what. But there's a *lot* of incidents that I do that I just go out, clear it up and I don't even report a lot of them because you're only making paperwork for yourself. If you can put it on paper someone comes back: 'What about so-and-so?' [In the old days] a lot of it wasn't reported. As long as you kept the public happy and there were no complaints —no-one could find any complaints to write into Head Office with—you were doing a good job. And I like to keep it that way still if I can, because you can make yourself a lot of work over some small incident sometimes. [His emphasis]

II. Control

In a job with a high degree of autonomy and unpredictability, the control of field staff assumes considerable importance, for 'the behavior of the members of an organization', as Blau (1963:2) observed with studied understatement, 'does not precisely correspond to its blueprint.'[4] Control is possible in an organization where staff can harbour ambitions about self-advancement allied with a desire for unruffled relationships with superiors upon whom they depend. The most explicit form of control is exerted through the field officer's immediate superior, the area supervisor, who occupies the key position of intermediary between senior officials based in offices in headquarters and district officers in the field. He is the line of communication between senior staff in headquarters and junior staff in the field, transmitting reports and results from the field upwards and instructions and decisions downwards. Though he has a good deal of contact with senior officials (and in the northern agency spends most of his time at headquarters) he has a strong loyalty to the field officers in his area,[5] retaining from his own earlier experience as a field officer what he feels to be a sympathetic understanding of practical problems. Despite the differences in conception of the job, field staff are linked in a strong 'web of solidarity' (Manning, 1977:144) which reaches across boundaries of age, background, or experience. An important task for the area officer, accordingly, is to represent and interpret the views of junior officials to seniors, and vice versa. As in all organizations, there is a marked interdependence between staff. Field officers rely on senior officials not simply for their own advancement, but for their support in disputes with dischargers and in cases which field staff wish to see taken to court. Headquarters staff are dependent upon the energy and diligence of their field officers in controlling and preventing pollution, in supplying adequate data and reports, and in representing the agency to the public. As go-between, the area supervisor has a sense of responsibility towards his seniors matching his sense of loyalty to his field staff.

In directing officers in his area, the supervisor monitors their work and performance and effects some control over their output. As the recipient of all their reports and assessments, formal and informal, the supervisor is in a position to exert his authority by moulding disparate practices and judgments in the field into some degree of consistency. 'Part of my job', said one, 'is to make sure that the approach in every district is similiar throughout the area and that we are consistent, in the same way that it's [my superior's] job to make sure that I'm consistent along with my colleagues in all the areas within the region.'

Supervisors can exploit a variety of opportunities for control of field staff. In the first place, they also know the job and the men in the field. Unlike most senior officials, the area men began their careers as field officers and worked their way up. Though they spend most of their time at their desks managing field work and consulting senior officials in headquarters, they are in frequent contact with their staff.[6] They know the difficulties, the frustrations, and the temptations of the work.

Secondly, area men are directly linked with their officers by the radio telephone in their cars. Field men can expect to be contacted not only with requests to handle complaints, instructions to attend pollutions or collect samples, but also on quite trivial matters. The radio telephone is only a partial link since often an officer cannot be located by it. He will frequently have to take samples or make visits which involve prolonged absences from his vehicle; besides, the terrain in some districts is such that reception of the radio signal is weak or non-existent. It is not unusual, then, for a call to an officer on the radio to go without response. But if repeated calls fail to produce an answer the officer may subsequently need to account for himself.

Indices of output are another opportunity for the control of field staff. Certain pieces of visible evidence of field officer activity serve as output measures, of which the most conspicuous is the sampling schedule which each has. The field man who has difficulty in meeting his programme can raise doubts about his industriousness. But it is often difficult to judge. So much of the job is unpredictable, and a pressing problem or a spate of pollutions or complaints must be given priority over routine sampling. Meeting sampling schedules will become more of a difficulty for field staff as they are faced with more demands from the organization. They must now recognize that laboratories need a steady flow of samples, and computers are increasingly relied upon for storage and handling of sample data.[7] Patterned sampling erodes the field man's personal discretion while it offers a more visible index of activity for the organization.

Senior staff in the northern agency did attempt during the research period systematically to evaluate their officers' work from their sampling records. Statistics were compiled of the number of samples (including formal samples) taken each month by each officer in each area. For a time the results were communicated to field staff at their regular monthly meetings. The analysis of sampling behaviour by headquarters, however, was immediately resented by field staff as an attempt to monitor their performance. It was not simply the principle of such an analysis being employed as a means of over-seeing their activities, they also objected to the 'league tables' (as they called them) on the grounds that they were an inaccurate and misleading data base for such an analysis, given the nature of the field man's work:

'... somebody may have one works [where] the idea is ... "We'll stat them twice weekly for the next four months". OK, but if that's on his cards to do so why should he be top of the table? It just means that he's——a lot of his time is taken up doing stats and it's——if he's doing that he's not doing something else. And stats do take a bit of time to do and you can write off a couple of hours if you're going to do it properly and make sure that you've got your i's dotted and your t's crossed.'

This experienced man's disquiet stemmed from the fact that the taking of a statutory sample, while visible in the organization, is (as he saw it) a potentially inefficient use of an officer's time, and the pressure to take a formal sample to show up well in the league tables placed an improper constraint on the discretion necessary to do the job efficiently. The stat, in short, is not necessarily the most appropriate way of securing compliance. But field staff fear that this might not be appreciated in head-quarters where senior officials might be unfamiliar with their routine strategies: 'One can say that because you haven't taken any stats in the last two years it's because you're a good lad, and people who you may have to take stats upon know that you're around a lot, and they know that if they're naughty——It's a two-edged thing, you can look at it two ways.' The existence of league tables was further evidence for older officers of inroads into their enforcement discretion and of lack of knowledge among their senior staff of the hard realities of pollution control work.

Another visible index of output in any organization, of course, is the production of paperwork. As more of the field officer's job is bureaucratized and made more 'scientific', he finds himself spending less time on the river bank. Instead, he must devote more time to drafting letters, making records, and writing reports in the office. Office work creates its own demands and its own routine, which area officers and senior staff can exploit as a means of assessing field staff's competence and industry. The power to evaluate is a power to control.

A final source of control over field staff stems from the supervisor's dependence upon the officer as gatekeeper—as supplier of reports and opinions. He may legitimately expect from his field officer detailed knowledge of and familiarity with any discharger or any 'problems'. The latter will reveal themselves to the area supervisor in sample results or—a more serious matter—in recurring complaints from the public or other water users about particular pollutions. Supervisors demand to be kept fully informed, and expect a field officer to have at his fingertips all relevant details about any pollution and the people involved. They look, in short, for a display of competence.

III. Covering

One of the paradoxes of pollution control work is that although they enjoy considerable autonomy, field men are acutely conscious of the scrutiny of their seniors and feel vulnerable to criticism. Like other enforcement agents they adapt by employing protective strategies. In effect an officer's decision behaviour in marginal cases where his work may become visible to his supervisor and where there is no 'obvious' course of action comes to be dominated by a desire to minimize potential damage to his position in the agency. 'I have to cover myself from my superiors,' said one, 'because they'll know I've been on site.' In practical terms it means taking that course of action which most reliably will avoid the possibility of criticism by their superiors. Covering is the response of officials enjoying substantial autonomy to efforts at organizational control. It is a constant concern: 'you have always to cover yourself.'[8]

Covering means 'doing it by the book'—erring on the side of caution and producing concrete evidence of activity when making discretionary decisions whose outcome will be known to the officer's supervisor or any senior staff. The result is that more rather than fewer cases are created, with staff conforming rather than innovating, for they realize that it is difficult for their superiors to complain of their being too efficient or too industrious.

The decision-making which has the potential for being known to senior staff, however, does not generally embrace the routine pollutions. The persistently troublesome problems and the large spot discharges are the pollutions which are likely to come to agency attention. In the most serious cases, a field man will always be concerned with covering himself in deciding whether or not to take a formal sample. His experience may tell him that it may be unnecessary or inappropriate in the circumstances, but failure to take one may become visible to headquarters.[9] The implications of such an omission may be serious, since legal action is rarely possible without the evidence of a 'stat'. Yet field officers often find it difficult to predict when senior staff may wish to consider prosecution, owing to the invisibility of the decision process in the higher reaches of the agency. Their sensitivity to criticism in this context was made clear by an officer of considerable experience:

> 'I've got to look over my shoulder, because there's always somebody that says "Well, why didn't you take a stat sample?" And this has always been levelled at you. I mean, you can go along and always think "Oh there's nothing to this", but then for some reason or other somebody has——a complaint's gone a big way round and ends up right at the top, and then, of course cascading down comes the "Why? Why? Why?". And somebody says "Why didn't you take a stat? You should have taken a stat." And you say, "Well, I didn't think it warranted." And they say "It's not for you to think." And there's this attitude that goes along, so some people will say "Oh I'll take stats", y'know, "regardless of whether I personally think [they] ought to be taken."

The problem is often particularly poignant since most serious pollutions are 'one-offs'—spot discharges, rather than persistent failures to comply—which deny field staff a second chance to 'get it in the bucket'. If a field officer who has taken a risk and opted to negotiate should subsequently have second thoughts, he may find his evidence has disappeared.

The practical problem of choosing a course of action is compounded by the difficulty of predicting the consequences of a particular pollution without spending a great deal of precious time on the problem:

> 'Somebody like [the agency head] isn't going to ... lay down the rules of where you should take a stat, or where you shouldn't take a stat. But the other way is slightly different you see. They can always be thrown out, they can't be obtained. ... And at times in retrospect you wish you had taken a certain stat, because it turns out that the whole thing has boiled up, completely beyond what you know it to be, but it's become political and the authority is being attacked and the chiefs can't defend it and say, "OK. We'll prosecute", because that silly fool down there didn't take a statutory sample. And in truth why didn't you? You know. Is it just that it's too much effort? Or should you spend a bit more time really finding out the extent of the pollution?"

One implication of this for field staff is that seniors find it much simpler to make negative rather than positive evaluations of their competence (cf. Niederhoffer, 1967:58). Their work is more likely to come to the attention of headquarters staff 'in practice ...', as a senior man put it, 'only when there's a cock-up'. An officer's doing the right thing at the right time is taken for granted. Failure to do so is much more noticeable. An officer who neglects to take a statutory sample which hindsight shows was necessary will find that his omission is more visible to senior staff than his taking a formal sample which is subsequently used in a prosecution by the authority. The commonest and most visible indices of field officer performance are negative, as an agency head explained: 'An area supervisor will sit there and say, "So and so has let me down again. He never did this, he never did that, I don't think we can take action, dearly as I would like to"—that's a negative evaluation. ... Very, very rarely does anybody come to you and say "Well done, that was a good job".'

IV. Competence

If it is much easier for senior staff to make negative rather than favourable assessments of his ability, the field officer is faced with an important practical problem of how he portrays himself to his seniors as good at his job. There are no ready measures to display as indices of success.

This is not a problem confronting policemen. In a sanctioning system, the enforcement agent can work by a penal control criterion of activity, as measured by arrests and clear-up rates (Reiss and Bordua, 1967:37; Rubinstein, 1973; Skolnick, 1966) even though police work actually involves little contact with 'criminals' and few arrests

(Manning, 1977:348). The policeman's indices of industry are visible, quantifiable, and relatively easy to produce. For instance, the traffic policeman who writes tickets for motoring offences may have a regular quota to meet (Petersen, 1971; Rubinstein, 1973). When in arrears he learns 'to sniff out the places where you can catch violators when you're running behind' (Skolnick, 1966:55).[10]

Performance measures which mean something to policemen are largely irrelevant in an enforcement system directed towards compliance. In fact the pollution control officer has as indices of activity neither the policeman's arrest rate nor the evidence which some other compliance system enforcement agents can exploit, such as the dollar amounts recovered by consumer protection officers (Silbey, 1978). Senior staff assume the pollution control officer will meet his sampling quota: to do so is no sign of hard work or competence. Since the formal procedures of the law are used so rarely, the number of legal samples or prosecutions in which an officer may be involved are never taken as a sign of competence. In a compliance system, indeed, resort to formal processes is often regarded as a sign of failure.

Other measures of field officer performance have to be used. Quantitative indices have considerable appeal for organizations, since they are easily handled and appear to be relatively unproblematic. But, as in other areas of enforcement activity, the quantitative measures available in pollution control have only limited utility since, like crime rates, statistics of 'pollution' and 'compliance' reflect the behaviour of enforcement agents rather than some 'objective' measurement of the phenomena themselves (Cicourel, 1968).

Sometimes a change of water quality is used as an indicator of effectiveness (see Wenner, 1971). 'Obviously the most apparent signs of success', said a supervisor,

> 'are signficant improvements in watercourses either because one has managed to get a discharger to cause his discharge to cease or to be brought up to a standard where a very signficant polluting input has stopped, and there is a corresponding improvement in the overall river system. You perhaps get restoration of fishery or enhancement of amenity or enhancement of the quality of the water such that it can be used for a far wider spectrum of requirements than it was before.'

Such measures, however, are of little immediate utility. Changes in water quality are rarely dramatic, but occur very gradually, and such a gross indicator rarely shows the direct impact of a field officer's activities. Even improvements in a number of effluents may not realize a noticeable change in water quality unless they constitute a substantial part of the flow of a watercourse. And there is always the problem of establishing a link between enforcement activity and its impact: most changes in water quality may be attributable to levels of economic activity and shifts in patterns of land-use, which have little to do with pollution control work. Only where a significant source of pollution stops suddenly—a large factory closing down, a new sewage treatment works being opened—will a striking improvement in water quality be discernible. The unfortunate irony for field staff, however, is that immediate changes in water quality are almost certainly changes for the worse which follow a major pollution incident.

The pollution control officer's nearest analogy to the police officer's arrest rate is a show of productivity measured in terms of paperwork, a feature assuming greater importance since reorganization. While paperwork is employed also as a protective strategy (cf. McCleary, 1978:145 ff.), field staff work on the assumption that the paper they produce and transmit to their superiors will be taken to indicate their industry and diligence. 'The reports you write, the information you may have' are taken into account. 'When [your supervisor] asks for a particular point, the more fully you can answer it [the better]. If your report makes good sense and it's well laid out [so much the better].' Paperwork is concrete evidence of work. Thus sample data and accounts of incidents become part of the raw material for an assessment of the competence of field staff. Reports fully and promptly written, and letters thoughtfully and sensibly drafted (thereby requiring little or no alteration when passed to the supervisor for approval and signature), all help create an impression of good work.

These indicators, however, suggest little of an officer's abilities in the field, where outputs are vaguer and defy quantification. The nature of pollution control work demands instead essentially qualitative judgments of field officers' abilities. Field staff are at an advantage in this connection, so far as routine cases are concerned, inherent in the nature of the job. Since the field officer is a loner, he controls his output to a substantial degree. His encounters are essentially private in most cases, for polluters do not often go over his head to his superiors, either to negotiate or to complain. As a result, supervisors are not often in a position to make first-hand assessments of an officer's skills in the field; they have to learn indirectly about his work, his negotiating abilities, the demands he makes, and his integrity. The officer's power to select what gets known to his supervisor may be exploited to support an image of competence which he can present in the reports he supplies.

Competence is a many-sided quality which reaches beyond those virtues of initiative and on-the-spot decision-making ability which are normally expected of field staff. Like policing, it is also a matter of flair, of the ability to be in the right place at the right time. One supervisor said he was happy when his people tripped somebody up:

> 'It's amazing how many people who, when you catch them with their trousers down and take a formal sample, will say, "It's been beautiful for the last two years and just this, just yesterday that this thing went wrong and your chap happened to come." And I, as far as I'm concerned, that's the sign of a good officer. When he goes on the one day in two years that something went wrong, that shows he's doing a good job.'

Serious pollutions, however, are relatively rare events. In routine work the field man must display himself as competent in other, subtler ways, showing himself to be on top of his job. While the primary virtue is the ability efficiently to get results, the dependence of senior staff on the man in the field for a continuous supply of information also places a premium on his knowledge. It is assumed that it is only the field man who 'really knows' what is happening 'out there'. Knowledge is again not a matter of understanding pollution control technology or the biological or chemical processes involved in water purification; such knowledge is taken for granted. A field man's practical knowledge demands instead thorough awareness of the nature and past history of any 'problem', including the record and results of prior attempts by the agency at control. An officer will be expected to know the personalities concerned, for this permits evaluation of the appropriateness of his action as well as prediction of their future behaviour. Ready answers to enquiries from above are essential. 'I don't like to say "I don't know" if [my supervisor] rings up and asks "What's going on at so-and-so?"' said one field man. 'I make sure I know what's happening.' (Cf. Wilson, 1978:83.)

Competence is also displayed in a detailed knowledge of territory. This aspect of the much prized composite quality of 'experience' takes time and work: 'You got to know the district. You got to know the people. ... I don't think you can do the job basically, in an office. Not a field officer's job. You got to hoof it. Know the rivers. I think I can say I walked the best part of the rivers in my district.' A supervisor can tap such first-hand knowledge without wasting time studying Ordnance Survey maps or other records, while the competent field man is in a position to handle a problem immediately in an emergency without having to research the questions of where a pollution may have come from, where it will go, or what damage it might do.

There was a call to say there had been an accident involving a lorry which had careered off the road spilling its load of drums, some of which were leaking an unidentified liquid. It had happened in Fred Wheeler's district—geographically the largest in the whole area, and an important catchment for potable supply—on a main road out in the country, somewhere between two towns about 15 miles apart. Wheeler happened to be in the office when the call came through. He told a senior colleague there was only one place on that particular fifteen-mile stretch of road where a lorry might be expected to run off it, at a bend where the camber of the road sloped to a ditch. Any liquid would gather in the ditch and there would be plenty of time to take preventive action, since it did not discharge to a stream for some distance. It would be easy, Wheeler said, to block the ditch off to trap the pollutant. Knowing the spot, he was sure he could handle the pollution on his own. After he had left the office his senior colleague observed with undisguised admiration how well Fred knew his district and how valuable it was for an officer to have such knowledge at his fingertips.

Wheeler phoned the officer an hour later to report that he had correctly pinpointed the location of the lorry and the likely consequences of the accident. Everything, he said, was under control. [From field notes.]

V. Postscript

Legal organizations, to be effective in attaining their mandate, have to transmit their policies to and make secure their control over the enforcement agent in the field. It is he who is in contact with pollution (or whatever problems cause concern) and his discretion which gives practical expression to regulatory policy. The control of junior staff is an enduring dilemma for bureaucracies, just as for staff the primary practical concern is how to succeed within the organization. Similarly, the autonomy which for an agency is a problem of organizational control is for the enforcement agent a source of personal control, providing satisfaction and professional pride.

The nature of pollution control work draws attention to these issues because it involves a substantial degree of discretion at field level among staff who are physically dispersed and isolated for most of the day while tending a complex and unpredictable environment. The work invites the imposition of organizational controls to minimize its 'centrifugal tendencies' (Kaufman, 1960). Organizational control is important for an enforcement bureaucracy not simply because agency effectiveness is at stake. Where the application (or, in one sense, non-application) of legal rules is concerned, there is an abiding concern for a notion of consistency of treatment.

It is difficult to maintain control where a scattered enforcement staff possess high discretion. Scrutiny by senior officials is easier in more serious cases because by definition they are more likely to be public knowledge in the agency. Where the privacy of routine cases provides an effective screen to scrutiny, however, the field officer is freed from the need to cover himself by acting according to his perception of the organization's expectations. These cases are the most numerous, and some have the potential for becoming 'serious'. The irony of organizational control in a compliance system is that it may impede performance and efficiency at field level, where enforcement is a matter of negotiation and bargaining demanding the flexibility of high discretion.

Indices of success in compliance systems are more elusive for field staff to attain than in sanctioning systems. In the latter, agents are organizationally rewarded by how well they measure up to the relatively concrete criteria provided by the enforcement process. The indicators of success in a compliance system, however, are vague because compliance itself is a vague and open-ended concept (ch. 6), and the enforcement agent has to fulfil a variety of roles demanding a wide range of skills and knowledge. He finds himself evaluated by the correspondingly vague notion of competence. In a compliance system, the 'good' officer is one who gets results quietly and efficiently.

5. Creating Cases

I. Working Definitions

The 'facts' of a pollution, like other forms of deviance, do not exist independently of the interpretative judgments which enforcement agents make about them. Any events or acts of possible 'pollution' have to be defined and assessed by a field man, having regard for their setting and context. The 'pollutions' produced as a result of these interpretative processes form the basis for any response by the enforcement agency.

Pollution control field staff have defined their central task in practice as securing the compliance of those who pollute, or risk polluting, watercourses. In utterly mundane terms it means getting polluters to do what they want them to do. But as with other types of deviance, pollution is not a simple, self-evident notion. Whether a given act or event is seen as pollution or not, to paraphrase Becker (1963:14), depends in part on the nature of the act or event and in part on what other people do about it. To understand how compliance is secured requires in the first place an appreciation of the means by which pollution may be defined, identified, and raised as a matter requiring the intervention of the pollution control agency. Certain kinds of events legally definable as 'pollution' are not followed by any action on the part of the field officer, supporting Becker's dictum that 'an infraction of a rule does not mean that others will respond as though this had happened' (1963:12). Since the legal rules about the control of water pollution are expressed through the discretion of field officers it is important to understand their working definitions of pollution in terms of the kinds of events, activities, or social settings which could give rise to action on their part, whatever its ultimate conclusion. The law itself, framed as it is in terms of strict liability, is not concerned with any niceties which might be provoked by extenuating circumstances. It simply proscribes discharging without consent or causing pollution (ch. 1, s.iv).

Identifying a discharge as 'polluting' is the first step in opening up the sequence of decisions to be made about modes of control. The possibility of pollution may often be established in the first instance by a third party—an onlooker or a user downstream—or by the discharger himself, rather than by the pollution control officer. When cast in this reactive role an officer inevitably becomes the recipient of prior definitions made by onlooker, complainant, or discharger (though he may, of course, subsequently redefine what has been reported). Since others do not possess the experience or diagnostic expertise of the field officer (except, in some cases, the discharger himself) and since reporting is contingent upon the knowledge that there exist means for 'doing something about the problem', it follows that the kinds of 'pollution' most frequently drawn to the agency's attention are those where the pollution or its consequences are easily noticed and identifiable as 'abnormal' or 'unusual'. The field officer will only discover examples of pollution himself, however, when he engages in proactive enforcement,[1] through regular surveillance by sampling or inspection or, if his suspicions are aroused, through spot checks. The chief means of proactive enforcement is routine sampling, which will bring to light or confirm the existence of persistent pollutions. But since field staff can sample in a day only a miniscule proportion of the discharges and watercourses in their districts, the coincidence of sampling when a serious 'one-off' pollution is occurring is very unlikely. Thus the potential 'pollutions' which the officer himself discovers are more likely to be viewed as 'marginal' or 'non-polluting' cases where no action may be taken.

Whether the 'pollution' comes to light as a result of proactive or reactive enforcement, however, the field man must decide at an early stage whether what he is observing constitutes a 'pollution' or not. Central to this question is a working construction of what is *tolerable*. This pliable conception structures the course of action taken. One part of this judgment is a *moral question* raised by the officer as to whether the case is one about which action ought not to be taken. The answer to this has two interrelated components. First, there is an assessment (sometimes negotiable) about whether it is *technically within the discharger's power* to do something about the 'problem'. It is widely believed, for example, that farmers contaminate watercourses every day in the normal course of their jobs; but, as an officer from a rural district put it, 'I wouldn't taken action if he couldn't do anything about it.' In other cases, dirty water is considered to be 'natural' or 'inevitable'. In urban areas field officers are well aware that heavy rain causes a substantial run-off of solids from buildings and roads into the nearest stream; those in mining areas know all about run-off from colliery tips in wet weather.[2] The second issue is more explicitly concerned with the discharger's *economic capacity to comply*, as gauged by common-sense assessment. The question is whether remedial action would impose an unacceptable economic burden. For instance, discharges frequently run from septic tanks in the grounds of country cottages, many of whose occupants are old people who have the sympathies of field men, since most of them are believed to be badly-off.[3] It is possible, technically, to control these discharges—but to take action here would be to involve the owners in an expenditure regarded as totally unjustifiable in the context of their limited means and the modest potential for damage. Field staff, by describing their inactivity as 'turning a blind eye', recognize that the formal law can embrace cases of this kind.[4]

The other part of the decision about what falls within the working definition of 'pollution' is a judgment as to whether the effluent will have an '*impact*'. If it will, it becomes a 'problem' and the officer must act; otherwise the field man will feel no professional obligation to take action. For practical purposes, therefore, 'pollution' is something which normally requires some action on the part of the enforcement officer and is defined not by legal rules but by the response of the enforcement authorities. It is an organizational and moral rather than a legal construct. This suggests that an event which the law might not recognize as pollution can in practice nevertheless be treated as such. For example, seepages often occur in urban or industrialized areas where waste materials have been dumped, often decades earlier. Water frequently accumulates on old factory and tip sites absorbing various kinds of powerful pollutants. From time to time the accumulation escapes by seepage through the ground and, most visibly, through earth banks around the tip. Seepages are usually treated by field officers as worthy of attention, therefore for practical purposes as pollution, since the pollutants involved are frequently visible and often toxic. Yet they know the legal position is less than clear. Some seepages can be sampled if there is a noticeable flow, but many are so insubstantial in quantity that sampling is not possible and it becomes difficult to talk of a 'discharge'.

Since the practical definition of pollution is tied to the notion of impact, it requires an assessment of setting, for setting determines the degree to which the field officer can contemplate tolerating the contamination. This assessment is of crucial importance. Because setting varies enormously, since each watercourse possesses its own unique characteristics, pollution is a highly relative notion. Some treated sewage effluents, for example, are cleaner than some rivers, while a 'bad' pollution on a potable supply river may not even be noticeable in an urban watercourse.

Two other features linked with judgments about impact are to be noted. The first, a matter of immediate significance, is an evaluation of the potential damaging effects of a pollution. The crucial issue here is whether or not the effluent is entering a watercourse which is used for water supply or amenity. Those rivers employed for potable supply create a special concern among field officers, reflected in a markedly greater willingness to define a discharge as polluting. This obtains *a fortiori* the nearer a discharge is to an intake for potable supply or some other use, such as fish farming, which requires water of very good quality. Since use of land for domestic and industrial purposes is largely incompatible with clean water, potable supply rivers tend to be found in rural settings where they also have considerable amenity value. Rivers in predominantly urban catchments, in contrast, receive large amounts of effluent treated to greater or lesser degree at sewage works, as well as discharges from industrial sites.[5] A watercourse viewed primarily as an effluent carrier will be thought better able to tolerate further pollution. An officer from a large rural district considered to be quite a 'difficult' patch observed, for instance, that he was 'sure the blokes in [the city nearby], where there are many polluting discharges into relatively fishless rivers, would not worry about things that'd cause great concern to Stewart and Fred' [two officers in very rural catchments with potable supply rivers].[6]

The second feature in assessing impact reflects the organizational context of discretion. This is a judgment about *noticeability*.[7] Many kinds of pollution provide visible evidence of their presence. Water is often discoloured. Steam indicates overheating. Oil is always particularly alarming since a very small quantity will quickly cover a large area of water with an unmistakeable iridescence. Other kinds of oil emulsify on contact with water, turning it, even when substantially diluted, the colour of milk. Pollutions which are high in suspended solids make water appear murky and leave deposits on the beds and banks of watercourses. Foaming water is evidence of the presence of detergent—though paradoxically the more water foams the cleaner it is, an important clue in industrial areas where river water almost always looks dirty. Noticeability does not, however, simply mean visibility. Polluted water often smells, and a serious and persistent pollution may turn a stream anaerobic, causing it to give off hydrogen sulphide, with its familiar odour of rotten eggs.

Contamination of water which is noticeable raises the field officer's anxiety because it indicates to the untutored that all is not well. It is important to bear in mind that, in contrast with trade effluents which disappear into sewers, most effluents discharging directly into watercourses are visible—and even if the discharge is not visible, its consequences (in the form of dead fish or other damage to the river's vital signs) certainly are. This places different constraints on pollution control officers.

Establishing noticeability requires attention to location and time, which in turn determine the significance assumed by such seemingly intrinsic features as the quantity and quality of the discharge and its continuity. Watercourses vary in size from the merest trickle in a ditch to large tidal rivers; size is significant since it controls the extent to which any pollution will be diluted, thereby more readily purified by natural processes, and less noticeable. Field men again adopt very different standards of tolerance, depending on the nature of their patch. Some spend their time inland on headwaters where streams are small and flows are low, while others work on predominantly tidal rivers where dilution is easily achieved and any pollution carried out to sea. The officer on a tidal river will be much more tolerant of pollution: one, for example, took a murky-looking sample of the discharge from a sand and gravel works which he described as 'a bit high' in suspended solids. 'If it was *really bad*,' he said, 'I'd go and have a word.' [My emphasis.]

Discharges may be permanent and persistent, or they may be episodic or isolated. Some will be regarded as 'temporary'. Many effluents which in other circumstances would be considered polluting—often highly polluting—are tolerated if the officer is satisfied that they will be relatively short-lived. Water pumped out of building excavations or road works, for example, may be permitted to flow directly into a neighbouring watercourse without a requirement that any solids settle out in the lagoons which are the typical method of purifying similar discharges more permanent in character. By the same token the continuous or intermittent discharge will concern an officer much more than one which is only occasionally made.

The amount of polluting matter discharged is also of considerable significance, but how significant depends on the capacity of the receiving watercourse. A discharge small in volume may be judged polluting if its watercourse is also small; if the stream is large, however, the effluent overlooked ('You take no action when your flow is piddling small') or, at least, considered insignificant and postponed indefinitely as a job to be tackled. 'That's not *real* pollution,' said an officer, noticing a small, discoloured discharge, and, by way of justifying his inaction, 'It's not "pollution" to put a little in. ... I don't see what harm that's doing to the [river] when there are thousands of millions of gallons a day coming down.' (His emphasis.) Weather and seasonal variations affect flows and, in turn, judgments as to whether a particular effluent is polluting. Summer flows, of course, are noticeably smaller, offer much less dilution and greater visibility to any discoloured water. In summmer, field men cannot afford to be too tolerant of pollution, especially with watercourses exposed to the greater scrutiny that comes with recreational use. Prolonged rainfall, on the other hand, provides massive dilution and little visibility, and in these circumstances the field officer will expect the first flush of urban run-off to bring oil, solids, and other matter washed from streets and buildings into his streams. Furthermore, the storm overflow system may well operate if the sewers are overloaded, discharging some of their contents directly into the rivers. The normal expectation of the officer in very wet weather, then, is that all watercourses will be swollen, turbid, and discoloured; any further pollution is less likely to be noticeable or do any particular damage.

What is 'noticeable' for the field man is thrown into relief by a conception of the 'normal' (Goffman, 1971). A particularistic knowledge of what is 'normal' for this watercourse, this discharge, in this weather, at this time of the year shapes an officer's expectations and equips him to attend to anything 'abnormal'. Is this pipe normally running? Is the discharge normally this colour? Does this treatment plant normally cause problems? This is all part of a search for incongruity which is a characteristic of enforcement behaviour in general (Rock, 1973a:177-9; Sacks, 1972).

Pollutions are rarely dramatic events in themselves, but are often more striking in their consequences. Any serious pollution is a hazard to the life which depends on the watercourse—all but the dirtiest of rivers support some form of fish and plant life. Bird and beast drink from them.[8] Fish floating dead in a river provide very conspicuous evidence of dirty water. Fish are sufficiently intolerant of pollution to occupy a position of immense significance in pollution control work: 'if fish are killed in even small numbers in rivers, they are a readily—observable sign of pollution which generates public concern.' (Unpubl. agency document.) Fish life is the commonest indicator of the general health of a river. For example, the presence of sticklebacks (fish known for their tolerance of dirty water) in a previously fishless stream is taken as an unambiguous sign of an improvement in water quality. The epithet most frequently applied to a dirty watercourse is 'fishless', while the aristocrat of freshwater fish is also the least tolerant of pollution, hence a watercourse of very good quality is usually referred to as 'a trout stream'. The trout stream, in particular, has a special significance for field staff and is a source of professional pride. Most are in country areas of high amenity value which attract outsiders—anglers and tourists—with decided views about what is meant by clean water. This inculcates in rural officers a notably greater intolerance of pollution in everyday work.[9]

Many pollutions will kill fish, or at least cause them distress. In either case they are readily visible to onlookers. A fish in distress swims erratically near the surface and readily suggests that it has something wrong with it. A dead fish floating on its side, exposing the silvery scales on the underside of its body, is even more conspicuous. Field officers assume that even the most ignorant onlooker will deduce that fish die because there is something wrong with the water.

The impact and noticeability of any contaminated discharge depend, of course, on the kinds of pollutant it contains. Organic pollutants take oxygen from water, and if present in sufficient quantities will kill fish and other life, including plant life. In a particularly noxious form, such as silage liquor or untreated sewage, the presence of organic pollution will be signalled with the growth of a grey fungus on the stream bed. Inorganic pollution is not harmful to aquatic life to the same extent, unless it happens to be toxic; thus its consequences are frequently less evident. Some inorganic pollutants, however, are intrinsically visible: a high concentration of inert solids, for instance, the waste from dyeing processes, or certain metals, like chrome, found in the effluents from metal-finishing works.

In general, the dirtier the receiving watercourse the less likely a discharge is to be defined as polluting, since the pollutant is less likely to have any measurable impact on the stream, less likely to be visible in already dirty water, and less likely to produce any noticeable ill-effects upon already scarce fish, plant, or animal life. From the field man's point of view, it is immaterial whether the water itself appears obviously polluted or whether a pollution has produced harmful consequences such as dead fish. In both cases it is open to others also to define the water as having something wrong with it: as one officer put it, having described an oil pollution which at the time was beyond his control: 'I prayed for rain and darkness'.

Formal consent standards attaching to a discharge may also be a criterion for action in a persistent pollution. Field men have ready access to consent documents and will treat as relevant the degree to which standards are exceeded. This is sometimes a reflection of the belief that if sample results show the discharge to be so much beyond the consent limits then the pollution must be having some effect on the stream. Almost certainly linked with this reasoning, however, is the desire to preserve the field man's credibility in the eyes of the discharger by demonstrably not tolerating a degree of pollution which may be seen as 'getting away with it'. The kind and amount of pollution which come to light in these circumstances are only knowable after a routine sample has been taken and analysed in an agency laboratory. 'Pollution' here takes on a much more precise and public character since sample results can be displayed and compared by field staff with the discharger's consent, whose parameters and limits mark out in a formal sense the degree to which the agency itself is prepared to tolerate pollution.

In deciding whether a breach of agreed standards is in itself cause for action, the field man will again set the pollution in the context of its location before judging the degree by which it exceeds the consented level. It must be remembered that the agencies, as a matter of policy, tolerate a certain amount of pollution partly in recognition of the vagaries of pollution control, and sampling and analytical error. Policy in the northern agency during the research called for compliance in about 70 percent of routine samples; with the review of consents, the agency has become seemingly less tolerant, calling for compliance in 95 percent of samples.[10]

Laboratory analyses often confirm a field man's suspicions or draw his attention to pollution previously unnoticed. In these circumstances staff then employ working norms which serve as rules-of-thumb to impose order on the judgment as to whether action is necessary.[11] An area officer with a mixed catchment described his approach:

'... you've got to look and see whether it's a big discharge or a small discharge. If it's a small discharge and say it's twice the limits, three times the limits, then it would be considered to be needing improvement. If it's a *small* discharge and it's just over the standard then it's *unfortunate*. If it's a *big* discharge and it's way outside the limits then it's *serious*—two to three times the limits. If it's a big discharge which is perhaps on a Royal Commission standard[12] and it is consistently turning out say a 30 BOD[13] and a 40 solids,[14] then it can be regarded as *harmful* and one would press for improvement to be carried out. ...

I will often say "Well this is twice what it should be, or three times what it should be," and decide accordingly as to whether it is serious or not.' [My emphasis.]

Three working categories are employed here: discharges which are 'small' and 'just over the standard' are defined as 'unfortunate'. No action is specified as the normal consequence of discovering this class of 'pollution'. One noticeably in excess of the standards (30 BOD rather than the 20 of the Royal Commission, 40 solids rather than 30) is'harmful' and requires 'improvement to be carried out'. A 'big discharge' which is 'way outside the limits' (by which is meant 'two to three times the limits') is described as 'serious' and presumably demands action, though no reference is made as to remedy. Though almost certainly more tolerant than the norms observed by the police in deciding whether or not to take action against speeding motorists,[15] these norms are somewhat less tolerant than those observed by some staff in areas of greater industrialization, as the following conversation with a field officer from such an area suggests:

FO ... I usually look upon a BOD over 100 miligrammes per litre as very bad.
KH Where the consent would be 20?
FO Where our normal limits are 20. The authority normal working standard is a 20, so I consider 100 miligrammes per litre to be bad—that's five times the BOD. Our pH[16] values are normally 5 to 9; I look upon anything above pH 11½ to be bad—that's from my experience in the laboratory that to get to a pH of 11 is easily done but to get to a pH of 12 you need a darn sight more alkali.
KH ... Do you have any other parameters?
FO Well, our non-ferrous metals limit has usually been one miligramme per litre; they have recently changed to 0.5 miligrammes per litre now. I look upon anything above 10 miligrammes per litre, non-ferrous metals, as bad. But we've got no set limits, I think it is all a personal judgment. ...

This final remark aptly sums-up the importance of context and individual discretion.

II. Discovery

Taking enforcement action requires in the first place techniques to bring deviance to light. This is a matter of *discovery*. Secondly, the deviance must be linked with an offender, a matter of *detection*. The distinction between discovery and detection is important. For example, the mobility of the evidence of pollution renders it potentially visible to a number of possible reporters but at the same time makes it more difficult to trace to its origin. Here, discovery is often easy, detection difficult (though aided by the fixed location of most dischargers).[17]

Field officers do not go about the business of bringing pollution to light in an unpatterned way. To do so would be thought ineffective. Field men know that pollution does not crop up at random and one of the arts of enforcement is to be in the right place at the right time—or at least to know where to look, for a particular problem in this form of enforcement is to forge a link between act or event and offender. Detection is less of a difficulty for certain other regulatory agents, such as housing inspectors (Mileski, 1971) or consumer protection officers (Cranston, 1979), whose enforcement domain is peopled with complainants. In water pollution, however, the evidence is mobile and there may be no complainants with a stake in reporting it. The mobility of pollution contrasts with other static forms of regulatory rule-breaking where enforcement is essentially proactive (e.g. Stjernquist, 1973; Mawby, 1979).

Since pollution is noticeable to others, and not necessarily at its source, it is suited to reactive methods of enforcement and field staff are accordingly sensitive to the need to respond to complaints. Mobility means that the deviance exists independently of its origin, allowing polluters opportunities of evading detection. In fact tying a pollution to its source can be a tricky piece of detective work. Ironically, however, the mobility of the evidence may increase the chances of discovery since it will be given greater exposure.

But field staff may have their suspicions and these can be translated into more frequent monitoring of the suspects in the hope of 'catching them at it'. A proactive programme of monitoring or inspection needs, in other words, to be informed by the experience of the officer: 'A lot of sampling is so unnecessary. ... When it comes to [discovering] pollution it's one in a thousand that you'll [do so] by sampling—it's in an [officer's] technique. ...'

There are three techniques which amount to working rules for the more efficient discovery of pollution. The first is '*know your district*', for knowledge of people and processes helps a field man organize his expectations and so inform his efforts. This is a major reason for the commitment in the job to thorough knowledge of the topography and geography of the district—the watercourse routes, the sites of potential pollution, and the location of sewage treatment works. Effluent in rural areas is usually discharged visibly into watercourses and can often be traced without much difficulty to its source. This all assists the job of detection, even though the officer has more terrain to police. The much smaller district of the urban field man, in contrast, has many more discharges per river mile, often from old-established industries in business for decades, more or less oblivious to changes in the law or the organization of water pollution control.[18] Effluent discharges are often made inconspicuous by buildings or the frequent disappearance of the watercourse into culverts. Even when a discharge is discovered, it is sometimes very difficult to establish where the other end of the pipe is: town plans or the other records which comprise the organizational memory are often incomplete or out of date, while physically tracing the effluent can involve the time-consuming and hazardous task of lifting manholes in the middle of city streets while filling sample buckets, proceeding by trial and error to narrow down the possible sources of pollution. It takes years, it is thought, to get to know an urban area well.

Knowing the area also means knowing the people well, especially those the officer will regularly encounter. After all, he is well-placed to make himself a familiar figure to the discharger:

'It's only really the chap on the ground who gets to know what's going on, gets chatting—and not just chatting with the management of the firm—chatting with the chap that runs the pretreatment plant, y'know, having a cup of tea with him and generally getting to know the individuals and the characters. ... Over the last ten years we've built up a pretty good network. ... I know what pubs to go into when I want to meet someone to have a chat. ...'

This familiarity assists reactive enforcement. It is an opportunity for the field man to educate people into what they can and cannot discharge into the river, and it contributes to a network of relationships which creates a means of learning about pollutions shortly after they occur. Familiarity can also encourage a mutual sense of trust which enhances the officer's capacity to detect and control because it encourages self-reporting of deviance.

Finally, it is also important for an officer to know about the processes which give rise to the particular effluents in his patch, since this helps the business of containment and on the spot diagnosis in an emergency. Familiarity with industrial processes also better equips an officer to handle encounters with dischargers, offering a means for judging the 'truth' of any account offered in explanation of a pollution. In short, he becomes more credible as a negotiator and as an enforcement agent.

Knowledge of geography, people and processes all contribute to the officer's 'experience', the virtue endorsed by the longer-serving officers. Experience is held to be the foundation for effective pollution control:

The stream looked normal to me, but the sample turned out to be very cloudy. Palmer had noticed that something was wrong with the water, even before taking the sample, and volunteered the opinion (subsequently confirmed) that cattle had been drinking in it upstream. 'No textbook can tell you this,' he said. 'It's experience. Years gone by in [one town] you could tell from the drainage whether it was a colliery or a pottery. You used to be able to go along to a pottery, say, and say, "What was going on here at about eight o'clock this morning?" ... You used to surprise some of them because they thought they'd done it before you'd notice.' [From field notes.]

The second working rule is '*be suspicious*'. It is not the suspiciousness prompted by regular contact with strangers, as in the case of policemen (Rubinstein, 1973); instead it springs from a recognition that, as a senior officer put it, '[a discharger's] priorities are not our priorities'. As enforcement of water pollution regulations places demands on polluters of time, manpower, and money, one of the basic premisses informing routine pollution control work is the assumption that every discharger has a substantial economic incentive to evade enforcement: 'Everyone will try to get out of it, or say it wasn't them.' Furthermore, it is generally assumed that it is an easy matter to pollute and get away with it. 'There are a hundred ways a firm can get rid of pollution into the river,' said an experienced officer, 'so long as they do it at the right time.'[19] The result is an imperative. 'You've got to be suspicious. You've got to be on your guard.'

Unless a watercourse is very small or the pollutant particularly powerful, the natural processes of dilution and purification will often make it very difficult for either proactive or reactive discovery to reveal the existence of pollution soon enough and in sufficient detail for it to be traced to a particular discharger. Officers respond by going immediately to the places suspected to be the most likely sources of the effluent. Suspicions are derived from past experience. As in other forms of enforcement, deviance which has taken place once is assumed capable of repetition. Or suspicion springs from an accumulation of complaints, from the physical signs of untidiness or carelessness, from characterization of a discharger from prior contact as 'the sort of person' who might be expected to give trouble, or characterization of a plant or business as 'the sort of firm or occupation' which causes problems. Farmers and small businesses (especially self-made men) frequently invite the latter response.[20] These are all examples of common-sense models of behaviour intended to predict continued misconduct. Suspicions may also be aroused in the course of routine inspection or sampling: where is the waste going? Why is that pipe discoloured? Why is there no sludge in the settlement pit? Hindered access to the effluent—any delay at the gate or other untoward behaviour which the field officer can interpret as stalling—may be taken as evidence that some sort of urgent remedial work provoked by his arrival, is under way. Since 'compliance' is an administrative definition and since production or treatment processes can constantly give rise to changes in water quality, field men must be ever-vigilant in the face of uncertainty.

Routine inspection of a watercourse is a constant search for the incongruous. In all of this the pollution control officer behaves much like other enforcement agents, of whom the policeman on the street is the best-known example (Rubinstein, 1973:218 ff.; Skolnick, 1966:48).[21] If policing pollution is not quite the 'uninterrupted sequence of suspicious scenes' (Werthman and Piliavin, 1967:56) which writers on the police describe so vividly, the pollution control officer must, even so, be continually on his guard:

'First of all you check your brook. If there's a change in vegetation, a change in the flow, a change in the colour ... you begin to be suspicious ... you get "nose". I sometimes head off to a certain place. ... I suppose sometimes I get a slight instinct for something happening. ... You need a "nose" for what's wrong. ... They call it "copper's nose". There's no substitute for regular surveillance all the time, no substitute for that. You can go to a stream fifty times in a year, but you've got to poke your head over the bridge—both sides—to be sure.'

Suspiciousness becomes almost an instinct, powerful and all-pervasive: 'When you go out on a Sunday afternoon [in the country], say, and you look over a bridge at the stream, you cannot help yourself. You're not looking at the stream, you're looking at the bloody quality and you can't stop yourself doing it.'

The third rule, which reflects the field man's position in the bureaucracy of pollution control, is '*cover yourself*' (ch. 4, s.iii). The field man's vulnerability to criticism from his superiors has implications for his decisions as to the circumstances in which action on his part is warranted. 'At the end of the day I'm responsible for the authority. So I have to take samples ... one's got to cover oneself.' In practical terms this concern is relevant only where supervising and senior staff in the agency may learn about a pollution; by definition these are cases which the agency regards as important, in particular, those which may have implications for the agency's public reputation. Decisions about action in routine pollutions are a matter for the field officer's discretion. His control in these cases is such that he alone decides whether or not to sample, whether or not to demand remedial action. The routine pollution, then, is one which the field officer can keep private as between himself, the complainant (if any), and the discharger (if he can be detected).

The pollutions which escape the field officer's shroud of privacy are those which are so noticeable that they attract widespread public attention, or those where a complainant is of such status that he may enjoy direct contact with senior staff. The agency itself has an important stake in avoiding complaints or in dealing with them efficiently. Pollutions which attract substantial publicity risk impugning the agency's competence. In fact the water authorities are especially sensitive to criticism in the media. Senior staff regard themselves as rarely far from the public eye, a view encouraged by the attention paid by some local newspapers to the water authorities. Indeed officials at all levels work on the assumption that the dominant public image of their agencies, thanks to a critical press, is of costly, overblown, publicly-funded organizations. Preventing cause for complaint, or making swift and effective intervention when grounds for complaint come to light, are opportunities for the agencies to demonstrate competence and efficiency to media and public alike. This concern is expressed at field level in a much more cautious approach: 'You take extra care where pollution could be visible—parks, lakes etc.—to the public. It could bring the authority's name into bad repute, so a greater watch is kept.' The guiding principle is one of minimizing the maximum damage. In practical terms this is translated into the maxim 'get it in the bucket', a rule of particular importance in 'one-off' rather than continuing pollutions, since there may only be one chance to sample. Besides, as field staff said, 'you can always throw it away later' if the pollution turns out to be unimportant and the officer wants to avoid the trouble of processing, that is, bureaucratically accounting for, his sample.

In general, wherever the officer believes a case to be one which may come to his supervisor's knowledge, it is appropriate to err on the side of caution, to 'go by the book' and protect himself from criticism. He becomes, in effect, more willing to define cases as 'pollutions' since the agency has grounds for complaint only when an officer fails to take action in an instance which officially warrants it—not the other way round.

III. Identification

Definitions of pollution are bound up with the process by which pollution is formally brought to light and identified. In contrast with some other forms of deviance, many kinds of pollution do not carry with them indicators which can be taken for granted by an enforcement officer (or anyone else) as unambiguous signs of their presence. Evidence decays. All but the most persistent kinds of pollution (now increasingly rare) are ephemeral and mobile. Isolated and episodic pollutions are soon flushed away. Even a persistent pollution will be less enduring given the right weather. Field men, as a result, are not only often in doubt as to whether they are observing a 'pollution', they have little time to consider what action should be taken. Where an officer wants to see what a new or unconsented discharge consists of, therefore, or where his own judgment suggests that a discharge may be polluting, he turns to his sample bucket, which offers the ultimate means for practical purposes of establishing the kind and degree of pollution.

Sampling takes a number of forms. The commonest is the routine sample of liquid drawn in a bucket from discharge or watercourse, transferred to storage jars and removed to the laboratory for analysis at the end of the day. A check of the water's temperature is often made on the spot and smaller samples may also be taken and treated immediately to enable a measurement of dissolved oxygen to be made, or to fix for the presence of certain substances such as cyanide. Mud and other solid matter taken from the beds of watercourses or from the ground over which a discharge has been made in the past may also be collected for analysis. Sometimes samples are taken of the natural life found in stream beds to construct a biotic index of the watercourse which will indicate its cleanliness by the kind of life the water will support. Such sampling generates quantitative data permitting trends to be charted which, if a deterioration is apparent, will draw attention to the existence of previously undetected pollution.

Sampling is one of the more tedious jobs. Some field men find the business of drawing the sample, filling the bottle, transferring its contents to storage jars, fixing for dissolved oxygen or cyanide, taking the temperature, and filling out sampling forms and labels all rather tiresome. If a discharge is small or inaccessible, sampling is also time-consuming. If an officer does not need to sample to fill his quota he may prefer to rely on his experienced eye instead and run the risk that an invisible pollutant like zinc is present. After all, as one said, 'you can usually tell if the water's good just by looking at it'. Besides, sample results are not instantly available, but take at least two or three weeks, sometimes longer, to come through from the agency's laboratories. In serious cases results can be produced more quickly, but in the case of organic pollutions the analysis itself takes five days.

Many officers routinely make an impressionistic assessment of quality, with a sniff of the water in the sample bottle or a sight of it against the light. The field man confronted with something which may be a 'problem' has to rely at the outset on his own personal judgment about the most appropriate action to take without waiting for sample results. In making this judgment he resorts to various clues which usually indicate some of the commoner forms of pollution:

FO ... you acquire the ... skill, with practice, of looking at a discharge and being able to say visually in many cases, that there's things like suspended solids—that it is going to be outside consent; it is clear to you that sample is going to be outside. So you're not messing around and wasting everybody's time. And even the organics ... I mean, inorganics are totally invisible, you can't really see inorganic pollution in general, but organics will colour more often than not. Farm effluents are khaki, milk is white, ... treated domestic sewage tends to have a sort of pale golden straw colour to it—depending on how dark the straw is it's a fairly clear indicator to you. And how turbid ... it then sort of goes through this darkening in turbidity. It tells you just about how much organic matter is present.
KH On the spot?
FO On the spot. And I mean over the years you can sort of pick a sample out and say 'It's pretty close, this one'; or, y'know, 'Do something about it', there and then, if you think there is anything that can be done. Or it can be very bad. Or you can pick it out and say 'That's a beautiful one, isn't it, there'll be no problem there,' and you look no further.
KH It's an art almost, rather than a science?
FO It's certainly not a science. ... [Take] temperature—you've got a thermometer in the car but you begin to feel what is about thirty degrees C. You sense that that's about thirty.

Identifying some kinds of pollutants, however, is simply not possible by rule of thumb assessments made on the spot. Indeed, some sorts of pollution may also for a time evade detection in the laboratory. Analysis can only reveal the presence of a pollutant for which tests are actually carried out and it adds to the costs[22] and time required (another constraint, of which field staff are well aware, against sampling too freely) to analyse routinely for pollutants on more than the usual parameters—BOD, suspended solids, ammonia. From time to time 'pollutions' with serious consequences occur which remain in effect undetected while laboratory analysis is directed towards testing for the presence of other substances. One such pollution which defied attempts at on the spot diagnosis and, for a time, laboratory analysis as well was described by a supervising officer:

'I can think of one firm that I think had tried to pull the wool over our eyes for many years. This particular firm had been causing fish mortalities in the stream—about half a dozen fish mortalities in the sixties. [Field staff] had been going along taking samples which had looked completely clear and satisfactory. And they'd reported that it was clear and satisfactory. In fact the effluent which was being discharged contained a very high quantity of zinc which was killing the fish. And it wasn't until we started analysing for zinc that we realized what the problem was. The firm had never said, "Well, we're discharging an effluent containing zinc." And the odd thing was the effluent looked better the worse it was. The clearer it was, the more zinc was in there. When there was less zinc it looked cloudy.'

The frequency and regularity of field officers' monthly quota of samples is controlled largely by laboratory capacity. Agency laboratories can process only a certain number of samples at any given time and organize the taking of samples to produce a steady and predictable flow. Laboratory constraints reach the field officer in the form of a curb against excessive or irregular sampling, suggesting the possibility that the agency's capacity to process data limits its ability to discover and identify deviance. For instance, an invisible pollution might not come readily to attention if an officer decides against sampling in the interests of easing the load on the laboratory: in the example above officers had reported that the water was 'clear and satisfactory'. But if there is any real doubt in marginal cases the officer's need to cover himself against the possibility of criticism from above normally resolves the issue in favour of sampling.

Sample results are a potential source of embarrassment to the agency, since, like criminal statistics, they are open to a variety of interpretations. When an intractable pollution occurs water authority staff frequently respond by increasing the sampling rate. Officers closely monitor those cases in which pollution may be attributable to a breakdown of treatment plant or some other apparently 'accidental' cause. This is partly as a means of enhancing control, permitting the field man to transmit his concern about the effluent to the discharger, the props of sampling conferring a certain sense of gravity and the act of sampling carrying with it the clear implication that the discharger is under scrutiny. Such sampling is also intended to produce better knowledge of the discharge, for then, as an officer said, 'you have facts to back up your statement'. An important consequence of these practices for the agency (connected with the familiar tendency to treat agency enforcement statistics as some sort of index of compliance) is that the publication of yearbooks containing sample results provides a statistical portrayal of non-compliance. A sampling rate stepped up to counter a 'problem' can place the agency and its staff in a rather poor light, suggesting inefficiency.[23]

Sample results are in effect the formal means of identifying the presence of pollution and are essential in any legal proceedings as a declaration of rule-breaking in kind and degree. The ultimate criterion for the presence of pollution in law, in contrast with other more familiar forms of offence, is a test carried out according to the supposedly 'objective' precepts of natural scientific (predominantly chemical) analysis. In private settings where the field man needs some seemingly objective, unambiguous evidence to support his demands for remedial action, sample results are also useful, though most dischargers are prepared to accept the field man's personal assessment of the nature of the pollution. In general, the presentation of sample results and the officer's interpretation of them assist in fostering an impression of scientific objectivity in the identification and measurement of pollution. In a large organization, especially, the data offer important evidence to senior management (otherwise insulated from routine pollution control work) of a problem which needs money spent on remedial action. Sample results can be portrayed as the outcome of laboratory analysis with all that that implies about accuracy, validity, and reliability, an impression doubtless enhanced by the use of computer print-out and the provision of results to one or more decimal places (cf. Sanders, 1977:206). Most officers have little difficulty in presenting themselves as specialists; their interpretation of the data is rarely challenged. 'They treat you as the expert,' said one experienced man. 'You are the expert, and you are expected to know.'

Identifying the presence of pollution by scientific methods is not the relatively simple and unambiguous matter it might seem, however. The data themselves are not unproblematic, for a certain amount of error is inherent in the very process of analysis and measurement.[24] For field staff, common sense takes precedence over their scientific inclinations. They are well aware of the amount of variation possible in 'scientifically produced' data and look with a certain scepticism upon laboratory results, especially when they diverge markedly from their own intuition or their expectations based on past knowledge. One field officer observed of the analytical results from a sewage treatment works, notorious in the area for its poor performance, that 'if you get anything below 80 [BOD] you start suspecting your own lab'.

Knowledge that sample results are not unproblematic indicators encourages some staff, when confronted with marginal cases where the pollution is not particularly noticeable, to adopt a degree of tolerance:

KH Why the greater tolerance would you say?
FO Well ... one factor must be analytical error in the laboratory, different laboratories. You can send the sample to 10 laboratories and get huge discrepancies in the results. I'm thinking particularly of BODs, perhaps. You can get variations. ... A report was done recently, I read, and the variation was something like 30 per cent discrepancy between the different laboratories in the same sample. So, that's one reason, analytical error, and

what they call sampling error.

Where the precise evidence of this form of rule-breaking is open to question, the deviant is often given the benefit of the doubt.

IV. The Organization of Discovery and Detection

Where an enforcement agency actively seeks out deviance by organizing its discovery systems to survey, monitor, or inspect, it mobilizes its resources *proactively*. Where outsiders, not under agency control, respond to rule-breaking which comes to light by reporting it to the enforcement agency, investigative and enforcement work is initiated *reactively*.[25] Both strategies are employed in pollution control work to enhance the discovery and detection of deviance.

Some forms of pollution are difficult to discover. Some are intended to escape discovery; some cannot be recognized as pollution; and some are events which are impossible to predict. Enforcement agents, accordingly, organize intelligence systems to circumvent these problems; in fact much of an officer's time is devoted to the discovery of pollution. The fact that the agencies have as their enforcement domain only a small segment of the population, and one almost wholly in fixed locations, means that a polluter is potentially knowable. Though theoretically any discharger is subject to the scrutiny of field men, enforcement activity is focused further and patterned to reflect a variety of assumptions held by agency staff about the likely location of pollution. Proactive enforcement is therefore selective, allowing the agency choice in the number and kind of cases handled, and informed by common sense models which predict the occurrence of rule-breaking.

As in other enforcement work, allocation of resources has direct implications for the nature and extent of the deviance discovered (Long, 1979). Similarly, different strategies of discovery adopted in the field generate different kinds of deviance. Proactive strategies are most useful in discovering the invisible, persistent, or episodic pollutions (Reiss, 1974). Reactive strategies, dependent upon complaints and tip-offs, are generally animated by the more conspicuous—hence the more serious—cases, many (perhaps most) of which are isolated incidents ('one-offs').

The presence of a complainant means that reactively organized enforcement is less private, since it is prompted by outsiders, and it permits agencies less opportunity to control their case-loads or allocation of resources. Deviance discovered by reactive means is relatively unpatterned as a result, in contrast with much proactive enforcement which is moulded by agents' predictions of where deviance is likely to be located.[26] This means that proactive strategies are particularly suited to those forms of deviance which are continuing or episodic—in terms of pollution, the persistent failures to comply. These are in general more likely to be regarded as 'less serious' matters.

Routine monitoring represents a compromise between surveillance focused on the potentially troublesome and the need to be prepared for rule-breaking from an unexpected quarter. Field men accordingly keep a check on water quality in general and the activities of certain dischargers in particular by regular sampling and surveillance of watercourses and effluents. They are largely free to organize their monitoring. They tend, however, to keep a closer watch on discharges which are potentially highly polluting or large in volume. A visual check of the general condition of the water or effluent is normally made when a sample is taken, and other inspections are often made without sampling. Proactive enforcement by inspection is often preferred to the time-consuming and troublesome business of 'dropping a bucket'. Inspection is a flexible means of monitoring, contact, and control. It makes the agency's continued presence known to dischargers without the trouble and cost of a sample. And since it provides the opportunity of personal contact with the discharger, it is a means of spreading advice or warnings.[27]

The routine sample is a monitor of the state of a watercourse or a particular effluent, intended to build up a profile of water and effluent quality. It also serves as an early warning of any deterioration in quality. The general rule is that the more important the discharge, in terms of its volume, contents, and the use to which the receiving watercourse is put, the more frequently it is sampled. Some large sewage works are sampled every day; major discharges may be sampled up to thirty times a year, while some minor ones are sampled only twice a year.

Regular monitoring is valuable to the field man for it assists the task of making enforcement appear routine. One discharger will find it difficult to complain about being the subject of excessive attention from the water authority—of being 'picked on'—because regular sampling means that all known discharges will be subjected to scrutiny. Besides, the ritual of sampling and inspection makes it easy for the field man to portray his activity as a normal part of his job, rather than as a search for evidence of deviance.

Routine river sampling is carried out according to a schedule arranged by the laboratory, standardized as to place and, as far as possible, time. In organizing their sampling and inspection of effluents, however, field staff adopt a different approach, informed by the imperative to be suspicious. An important assumption shapes enforcement practice here, namely that deterioration in water quality will frequently come about by design rather than accident. It is believed that many dischargers, particularly industrialists, have such control over their manufacturing and treatment processes that they are able to alter the quality and quantity of their effluents virtually at will, as economic considerations dictate.

The image of deviance as controllable moulds the form of proactive discovery adopted. Officers must be prepared to do the unexpected to pre-empt the possibility that a polluter will effect a short-lived remedy for their benefit. Thus routine sampling of effluents, ordered though it may be by laboratory capacity, must be organized by the field officer to appear unpredictable.[28] Any regularity in monitoring activities allows the discharger, in anticipation of a sample being taken, an opportunity to step up his pollution control to produce a better effluent. 'It's worth money to a firm', said a supervisor, 'to know when we are coming to sample.' Unpredictability, however, is difficult to achieve in practice. The tedium of routine sampling means that a conscious effort is needed to do the unexpected: 'randomness isn't an in-built thing, you always have to think about being random.' Other demands on officers' time tend to impose a pattern on their behaviour which is difficult to break without substantial disruption to their schedule and mobility. To avoid excessive travelling, they usually organize their sampling runs to cover a particular area. Or they may take advantage of visits to dischargers for other purposes to fill a sample bucket at the same time. Some sample at times which fit in with their journeys to and from home or the laboratory. Some like to avoid the worst of city centre traffic, while those who work on tidal rivers have to time their sampling to coincide with low water, since many discharges are only accessible then. It is difficult to disguise the routine nature of proactive sampling or to resist the temptation of dispatching the task more swiftly: 'The trouble with this job is you get into habits. You go down there on Tuesday because you can park.' There is only limited opportunity, in effect, to break the patterned nature of their work if field men are to do the other things required of them.

But some attempt at disguise is essential. Hence staff sample covertly where they can, without gaining access to the discharger's premises; or they sample at night or weekends when they are not expected to be at work, and any routine sampling done during the day is never by appointment:

The settlement pit in a sand and gravel works was well stirred up and discharging large amounts of solids into the river. 'I'll go and have a word with him', Lawton said, 'but I won't push it unless he gets stroppy.' We went into the manager's office. Lawton adopted a stern attitude: 'About your settlement pit. It's really not good enough. When did you last clean it out? I don't want to take a sample, but it's not good enough.' The manager replied it was cleaned out that morning, and came with us to take another look. Lawton ended the conversation saying he would be back 'tomorrow' [Friday] to see if it had been put right. 'I never come back when I say I will,' Lawton muttered to me, walking out of the works, 'I might look in on Monday.'[29]

Even if an officer is working on a pollution problem requiring repeated visits, he continues to do the unpredictable wherever possible to provide the polluter with few opportunities to organize his activities so as to create a spurious impression that things are under control.

'Pure' proactive enforcement (unguided, that is, by procedures for inferring a potential for deviance) is rarely engaged in. There are too many demands on a field officer's time, and discharges are too numerous and too scattered for him to spend the day surveying dischargers in the hope of observing pollution actually occurring. For this reason time is devoted to monitoring of a much more focused and patterned kind in which officers employ techniques to predict the sources of pollution, actual or potential.[30]

This is the art of inspection. Special skills are needed to read the signs of dirty water. The officer must examine the colour, contents, and flow of the watercourse, and the condition of vegetation, trees, bed, and bank for the existence of deposits, stains, or fungus. The banks or river walls must be scanned for obscure discharge points. The colour of a pipe's interior, fresh deposits in the bottom of a pipe or stream bed, can all yield clues. The older men in particular portray inspection as an art whose techniques, absorbed over a long period, become almost intuitive. These technical skills have to be allied, however, with social skills to enable an enforcement agent to predict the likely source of trouble.

Expectations generated by the polluter's occupation, past behaviour, or performance mean 'you generally know where to look for trouble'. 'If you've got a company that consistently, since you can remember, has turned out a good quality effluent,' said a supervisor, 'obviously you do look at them from time to time. But you're wasting your time looking at them when [a firm you know to be bad] up the road might be doing something terrible.' Past performance is particularly important; indeed, some dischargers are identified in the organization by reference to their past pollutions. This organizational history generates expectations upon which surveillance can be premissed, even though no action may be possible at the time. 'Somebody rang up and said "There are fish in distress in the river. ...", said one officer.

'We went along, took samples of the river water and found there was zinc present, and we tracked it back. And we knew very well it came from a particular firm. But we were never able to prove it. And of course from that time on you pay more attention to that particular firm. In other words you build up a history.'

Expectations also inform decisions about 'purges'. In at least one region enforcement is occasionally organized and directed against particular targets, in much the same way as police forces occasionally indulge in crack-downs. 'There was a time when, about eight years ago,' recalled a supervisor, 'one did have a purge, and one still has purges, but now one has district purges against perhaps particular industries or different offences. ...'

Inspection can also be made tactically useful. The lack of a distinctive appearance as an enforcement agent confers an unobtrusiveness to be exploited when it is important to examine a discharge covertly. But the field man has a choice. Inspection can instead be employed as a subtle, yet conspicuous, warning to a discharger that he is under surveillance where it is tactically in the officer's interests to give a visible reminder of his constant scrutiny (cf. Cranston, 1979:74). This style of proactive enforcement is encouraged by constraints against excessive sampling and employed for deterrent purposes, where a discharger is thought to be contemplating misconduct. 'When they see this yellow Escort,' said one of the most experienced men, 'they know Alan Armstrong's around'. The field man makes sure his presence is noted:

We went to look at an urban stream where a small engineering firm had been causing problems. Griffith had been there only a couple of days earlier. He parked conspicuously outside the factory's main entrance and went ostentatiously to look at the stream from the bridge nearby. Griffith thought the water in reasonable condition, but was more concerned that the factory foreman had seen his car and had noticed his continued interest in the stream. It was all part of the tactic, Griffith said later, of keeping people on their toes.

Conspicuous inspection is a useful move with those dischargers judged to be more responsive to a discreet deterrence by way of an amiable expression of interest and gentle advice as a means of displaying that the officer means business: 'Just poke your nose in. A social call, once in six months, just to show you're still around. My theory is to let them know you're still watching them.'

Sometimes it is essential to monitor a polluter's progress while not endangering delicate personal relationships. This requires entirely covert inspection, which is only possible where an officer does not need access to the discharger's land to be able to inspect the effluent or take a sample. On the other hand, when dealing with the unco-operative, the field man prepared to play a waiting game can catch out the discharger by being on hand to sample if the chance arises. This is the tactic to adopt when an officer wants to take action, but cannot:

> Andrews was having trouble with a garage which regularly discharged into the canal a small but highly polluting waste containing a noxious chemical from car wash water. The garage owners were proving to be unco-operative, having failed to respond to a letter threatening action Andrews had sent. The last sample he had taken was 600 BOD, but he was unwilling to take another because the discharge was so small. He decided to adopt a tactic he described as 'waiting for something to happen'. 'I'm not going to spend half an hour with a bucket, hanging over the edge of the canal and probably falling in. ... I'll just keep coming back. ... And, you know one of these days it'll start raining ... and flush the contaminated water out of the system' [thereby creating a polluting discharge].

Proactive enforcement is also valuable in giving an officer access to premises and the opportunity to extend the ambit of his pollution control work. When on site for monitoring purposes, the field man can use the occasion to look around in search of evidence of other pollution or potential sources of trouble, in a manner reminiscent of the police practice of 'checking someone out' (Rubin, 1972).[31]

When enforcement in pollution control is reactively initiated, descriptions and evaluations are provided in the first place for the agency by the complainant or onlooker who reports. The agencies' sensitivity to noticeable pollutions is reflected in the emphasis given to reactively-organized enforcement by efficient communications and other forms of organizational intelligence. The officer in the field on routine work is equipped to respond immediately to emergencies, being kept in touch with his area office by radio telephone. At least one field officer is available, according to the dictates of the duty-rota, to take incoming calls; and at weekends another rota of field staff on stand-by at home provides cover when the area office is unmanned.

Routine monitoring plays a relatively small part, not unnaturally, in reactive enforcement. Sewage treatment works and water supply intakes constantly check on the quality of their inputs, however, and this monitoring occasionally succeeds in drawing attention to serious pollution.[32] In the nature of things, however, only the measurably abnormal will come to attention, and the information which can be given is inevitably unspecific. But at least the officer can sometimes pick up sufficient information to allow him to generate a plausible suspicion of the source of the pollution, and to direct his enforcement activities accordingly.

In general, it is the more important cases of pollution which are discovered reactively since they come to agency attention in circumstances where usually there is very strong evidence of pollution. The most noticeable pollutions tend to be discrete incidents ('one-offs'). These are often 'serious' matters and are particularly given to reactive enforcement. Here the agency must inevitably rely upon the informant who effectively takes the first step in initiating enforcement. The informant may often be the 'victim' of pollution if he is downstream and his use of the water is impaired or stopped, while the onlooker who reports a pollution is usually someone who has knowledge of the existence of an organization for pollution control, and whose enjoyment of water as an amenity is threatened.

Reports come from many quarters, and can range from a message from a passer-by of something unusual in the river to vigorous complaints by a downstream user. With such public non-compliance, any complaint is treated as a potentially serious matter by field staff. It is difficult for them to keep such matters private, since each complaint made to the area office is supposed to be routinely recorded on a form which is filed in the office, with a copy to head office. In practice, however, as a senior officer put it, 'We ignore some of the trivial ones. ... The number you do's nothing like the number you put in.' Complaints which come into the office and are logged may no longer be handled in private, officially unreported negotiations conducted by the field man. Instead they demand action from him because they open the way for a negative assessment of his competence. Field officers appreciate that a complainant who is a water user has a stake in having his complaint attended to: 'if it appears to complainers that justice isn't being done, it could cause problems for you.' Since complainants invite attention from senior staff they take priority over other obligations. They also spell trouble in the form of more work, especially if the complainant is persistent and demanding, or the pollution provokes numerous reports.

Furthermore, the reactive role of the officer as recipient of an account made by a third party means reliance on the descriptive powers of that person. Complainants are regarded as untrustworthy sources of information: untrained eyes are prone to inaccuracy and many succumb to the temptation of exaggeration. Field officers are well aware that while they are 'repeat players' in the business of pollution control, and that all but the most serious cases are routine events from them, pollutions are unusual, possibly dramatic, incidents for others. Field staff's distrust of lay descriptions means not that their seriousness may be discounted, but that each one must be treated as potentially serious, to guard against the occasional pollution which is serious. This strengthens the officer's conception of 'trouble' associated with complaints. Anglers are believed responsible for many of the extravagant reports. They are numerous and well-placed to observe pollution. Their interests are represented by anglers associations regarded in the agency as ever-vigilant, vocal, well-organized pressure groups. Attending to problems raised by anglers is a time-consuming business which disrupts more important work, leading to a cynicism and lack of sympathy among agency staff in general. Anglers are often the butt for office jokes: reports of 'pollution' or 'dead fish' which upon investigation seem to be without foundation will be dismissed as the work of 'a typical angler'.[33]

Complaints also threaten trouble for the agency. The involvement of a third party can make any possible pollution a public matter in which concern about the efficiency of the authority and its responsiveness may be raised, a matter of considerable significance for an agency operating in a political environment marked by considerable ambivalence. A complaint is a chance for public relations work, an opportunity for the agency to demonstrate its worth making the source of a complaint a significant matter. Agency vulnerability to public criticism leads them to be more responsive to complaints raised by those in positions of power: 'if you get a complaint brought in by an MP you tend to devote more time to it', said a supervisor. 'It might be seen as an unreasonable attitude, but that's the way it is.'[34] This vulnerability to potential public criticism reaches down to field officer level in the form of a rule to take all complaints—at the outset—seriously: 'from the authority's image point of view,' said a supervisor, 'it's got to be that way.' (Cf. Black, 1971:1095; Cranston, 1979). Indeed, for many field men the working definition of a 'serious' pollution is 'basically anything that's going to cause a great amount of public reaction'. The implication of receiving a complaint, then, is that the field man has to treat it as 'something to act upon' (cf. Emerson, 1969:86), tiresome in practice as it often is: 'If anyone complains about anything, we investigate it ... [but] so many are trivial or are cleared up by the time you've got there.' Where a third party is not involved, however, the field man is free to handle an incident as he pleases and his position as a repeat player leads to a narrower conception of polluting incidents as those potentially noticeable matters which may prove 'troublesome' or 'difficult' (cf. Emerson, 1969:84).

A complaint, however, can be turned to advantage by the field man who needs access to premises which he suspects to be the cause of the problem. The way he presents himself in these circumstances is an important matter, given that the frequent consequence of a complaint is a request to the discharger to spend money on a remedy. To present the request for access and co-operation in the context of a complaint made by another user downstream is to give the discharger less opportunity for objection. The field officer, in other words, is able to smooth the task of enforcement by presenting himself innocently, as someone responding to the urging of an unknown third party (the identity of a complainant, if known, never being revealed), who only interferes because he has a job to do.

Reactive enforcement poses a paradox for field staff. While the mobility of pollution increases the chances of discovery, it often makes detection more difficult. Unless the polluter himself reports a discharge of effluent, reactive discovery means that a pollution has already become public. To forestall this difficulty, officers, like other enforcement agents (Skolnick, 1966:120 ff.; Walsh, 1972; Wilson, 1963) extend their intelligence network by cultivating informants who will report an incident to the agency as soon as they learn of it. The practice is widespread and carefully pursued by field staff. Those, like anglers or downstream users, with vested interests in clean water make good informants. Others sometimes live in houses strategically placed on river banks. 'I've educated Jim the postmaster [who lived by the river] so that as soon as he sees something,' said one officer, 'if he's got the slightest doubt, to call me.'

Responsiveness to complaints adds to the network of informants or observers. The most important informants are those insiders in factories which themselves cause pollution. Thus the 'keen angler in the factory's the man you chat up'. Some are believed to inform because they want their employers to spend money on pollution control; some because they bear a grudge (thus endorsing the impression that the activities of pollution control staff cause 'trouble' to dischargers). The informant located on the premises is more useful than the river bank observer since the field officer can be clear about the origin and nature of the pollution. The officer possesses a confidence based on inside knowledge which allows him to take action, even though the effluent flow may have already ceased: '... by the time we got there, there was nothing coming out of the pipe. But you knew who it was and you could go in and chew a strip off them.'

Cultivating informants is made easier where larger concerns are involved since they are comprised of individuals with differing degrees of commitment to the interests of their organization. Indeed many employees in industry are regarded as impatient of the sometimes modest efforts their employers make to clean up their effluents. Informants are recruited on routine sampling visits and encouraged by field staff to report any untoward incident or abnormal effluent: 'You build up a series of contacts or spies who aren't particularly loyal to the factory who ... when you go into a factory will run over to you and say "You know what they were doing last night? They were emptying their acid baths." ' Since many pollutions carry with them few signs about their origins, such reliance on inside information is an effective way of producing knowledge about offenders.

For these important reasons field staff make special efforts to encourage dischargers always to get in touch with them when they have any problems at all; indeed, a significant index of the 'co-operativeness' of a polluter is the extent to which he is prepared to report his own deviance. This is another reason for officers' attempts to maintain friendly relationships wherever possible by displaying 'understanding' of problems, and by inculcating the feeling that dischargers can 'trust' the officer. In practical terms this means that pollution voluntarily reported will not normally be penalized.[35] A major rationale for the maintenance of good relationships between enforcement agents and their clientele, then, is that as well as assisting case-working with the deviant it enhances the discovery and detection of deviance.

V. Postscript

Pollution is the ephemeral result of an act or pattern of deviance posing problems of discovery and detection (though the difficulties of transience and a shifting territorial location are even more acute in the control of air pollution). The nature of water pollution is extremely variable, with implications for its noticeability as a matter for enforcement. Some pollution is evanescent and easily dissipated; some is readily noticeable; some is enduring and predictable. But all pollution of watercourses is mobile, and there is a limit to how long a slug of dirty water will persist before dissipating or purifying by natural processes.

Of crucial importance in defining the existence of a pollution and determining the enforcement action to be taken is the judgment of field staff. Deviance in theory open to a control response is always screened at field level for access to the system of control by officers who decide which cases come to agency attention for designation as 'pollutions'. The field officer's conception of seriousness is bound up with his vulnerability to organizational control, whereas his agency's conception of seriousness is linked with a notion of publicity. In using his discretion as to whether or not a pollution exists, the field officer is implicitly designing his (and his agency's) case-load of problems about which action is to be taken. Implicitly also, he is expressing agency policy about pollution control through the cumulative effect of discrete decisions, making concrete the aspirations and rules embodied in the law and marking out yet another series of boundaries between the legal and the deviant. In pollution control work, however, definitions of deviance may also be bound up with technological capacity to identify rule-breaking.

All enforcement work is adaptive. Staff have to develop various systems of intelligence to facilitate the discovery and prevention of deviance, much of which, of course, is intended to defy discovery. Enforcement is accordingly organized to make pollution quickly discoverable. Where it is persistent or episodic the agency can employ proactive techniques of monitoring and surveillance; where it is an infrequent or isolated event, resources must be effectively organized to permit a swift reactive response. With proactive enforcement, the selection and coverage of 'problems' to be handled as 'pollution' is largely within agency control. Pollutions arising by outside report and complaint, however, provide sources of deviance whose distribution and intensity are largely outside agency control. With both proactive and reactive strategies the expectations of enforcement agents inform their judgment as to where the source of the rule-breaking is likely to be located. The predictability of deviance is important, for both rule-breaking and rule-enforcement are patterned, not random, processes.

Pollution is not self-evident. It does not define itself: enforcement agents do. Dirty water is transformed in a sometimes subtle and complex process into a 'pollution' in the creation of a case which on moral and organizational grounds is worth working. In this transformation it is setting, not rule breaking *per se*, which shapes an enforcement response. The central feature in determining the reach of the pollution control process is not so much a question of what 'pollution' is, but rather what makes a 'case'.

III. Securing Compliance

In the next three chapters, attention shifts to the nature and content of enforcement work in pollution control, for in a compliance system agents not only create cases, they also work cases. What field officers understand in practice by 'compliance' is discussed in Chapter 6. The analysis suggests how enforcement strategy is linked with conceptions of deviant and deviance. Bargaining is identified as the central feature in the game-like compliance process. The array of tactical options and the wide variety of enforcement moves employed in a compliance strategy are the subjects of Chapter 7. One of the book's themes, the pervasiveness of moral judgment in the enforcement of law, is made explicit in Chapter 8, which shows how heavily officials rely on moral evaluations of the events and people they regulate, even though they enforce a strict liability law.

6. Compliance Strategy

I. Compliance

Enforcement behaviour in pollution control is determined by the play of two interconnected features: the nature of the deviance confronted and a judgment of its wilfulness or avoidability. Field staff draw a distinction in practice between deviance which is continuous or episodic *('persistent failures to comply')* and that which consists of isolated, discrete incidents *('one-offs')*. Because the persistent failure to comply is open-ended deviance, it is more amenable to detection. The one-off, on the other hand, is an unexpected discharge of relatively short duration, and hence less open to detection. It may be accidentally caused, but its unpredictability carries with it hints of attempts to evade detection.

Detectability is linked with avoidability. The field officer's common sense suggests that where a pollution is persistent, its perpetrator is easily detectable. Yet this ready detectability only rarely suggests wilful persistence. Instead the persistent failure to comply prima facie implies that the rule-breaking is unavoidable, the result of inadequate resources or knowledge. Where deviance is unavoidable, a strategy of compliance is called for to repair problems.

A one-off is a more sinister matter. The intimation of possible wilfulness, the suggestion that a pollution is preventable, prompts a penal response: there is no problem to repair, simply rule-breaking to punish. The one-off pollution, in other words, has the potential for resembling a traditional criminal act, making a conciliatory style of enforcement inappropriate, for there is a suspicion of 'trying to get away with it'. Its limited duration means that, compared with the persistent failure to comply, it is more difficult to discover the deviance, more difficult to detect an offender, and more difficult to establish that offender's motivation. Discerning cause here acquires considerable significance, since it is only a judgment about blameworthiness which distinguishes a preventable one-off from an 'accident'. Whether a pollution is regarded as 'accidental' or as something more ominous depends upon an image of the polluter. Such imagery 'explains' pollution.

Yet this is not to suggest that a penal response is never made to persistent failures to comply. Where an agency prosecutes in these cases, however, it is for a failure of the compliance process, a sanction for the irretrievable breakdown of negotiations.[1] Here, the failure of the compliance process suggests a persistence in wrong-doing whose wilfulness aligns the case with a deliberate one-off.

A notion of efficiency is the field man's major concern. The expeditious attainment of his given objectives means securing compliance at least cost to his future relationships with the polluter. With a persistent failure to comply, he must clean up a regularly dirty effluent. And though the one-off can prompt a punitive response, a concern for efficiency is expressed in the preventive work done to forestall recurrence—whether or not the incident is judged 'accidental'. In both cases the active co-operation of the polluter is required for the success of an enforcement strategy aimed at conformity.

What an enforcement agent understands by 'compliance' depends on the nature of the problem he is regulating. Persistent failures to comply are the episodic or continuing pollutions which are frequently or consistently above consent limits, either owing to inefficient treatment plant or inadequate attention to pollution control. The nature of this deviance is open-ended and persisting. It often comes to agency notice through proactive enforcement by routine monitoring and may have long been going on and only recently discovered or recently defined as pollution. Such pollutions are typically not regarded as 'serious' (most 'serious' persistent failures to comply have now been cleaned up), though field men may be anxious about their cumulative impact; indeed they are a classic embodiment of deviance as a state of affairs rather than as an act. The normal control response is to prescribe remedial measures to be applied in the course of the continuing relationship between officer and polluter.

Field staff sometimes describe persistent failures to comply as 'technical' as opposed to 'drastic' pollutions, a term reserved for serious 'one-off' cases.[2] The use of the word 'technical' aptly indicates a discharge above consent to which blameworthiness does not attach. Most such deviance is treated as routine, uneventful: 'I think people see discharges as outside consent conditions, and have been for a long time, and people don't take that much notice'. The enforcement response is modest; 'We would', said an area supervisor, 'just tend to niggle away at the bloke.' A persistent failure to comply will be treated as a more serious matter only where the discharge is regarded as substantially beyond consent limits and the pollution is noticeable.

The responses of the persistent polluter are, however, carefully monitored by field men. Here enforcement is a continuing, adaptive process. Persistence allows the construction of a career assembling past problems and past efforts at remedy made by the agency, interlinked with the polluter's responses to them.[3] To think in terms of an enforcement career is a useful device for the officer, structuring his expectations and moulding his responses. Present conduct has meaning and significance conferred upon it by past history; career creates context.

The one-off pollution is potentially of much greater significance. Almost by definition it will be serious simply because it is less open than a continuing pollution to detection through routine monitoring. The one-off pollution is a discrete event; it may be momentary or longer-lasting, but it is bounded in time with a finite beginning and end (its end sometimes the result of enforcement work). It is critical rather than chronic deviance; an 'incident', where a persistent failure to comply is a 'problem'.

The one-off normally comes to agency attention as a result of complaint from a third party; sometimes, though, it is difficult to detect: a discharge can be turned off or an effluent diluted before a field man arrives on the scene. And where a persistent pollution has overtones of the (for now) 'unavoidable', or possibly the 'careless', the one-off prompts a different moral categorization. It may be 'accidental'—a spillage, a leakage, a breakdown of treatment plant—or, more ominously, 'deliberate'—acid baths being emptied or settlement tanks flushed out.

Enforcement activity with one-offs is directed towards correcting damage done, preventing its recurrence, and deterring others. Because the one-off pollution by definition implies the existence of a discharge which is not normally made, or the presence of a pollutant in a normally clean discharge, compliance is relatively simple to secure where the effluent is still running. Field staff will demand a stop to it, and the more serious the pollution appears to be, the sooner they will expect effective remedial action. Compliance here can be instant: taps can be turned off, an effluent pipe blocked off, polluting liquid diverted or disposed of in waste tankers.[4]

Where persistent failures to comply or preventive work with one-offs are concerned, compliance is a continuing process, an organized sequence of requests and demands placed upon the regulated. Compliance here is elastic, and depends on the apparent 'progress' made by the polluter and other exigencies affecting his relationship with the field officer. The issue is not usually the discrete one of whether or not the discharger will take action, so much as whether he will act as quickly as the officer wishes and to the extent required. The stance adopted by the field man during negotiations and the sorts of demands made of the polluter will be influenced as much by interpretations placed upon the polluter's response as by the nature of the 'problem' to be remedied and the resources available for tackling it. The technical implications of the behaviour addressed by law require this fluid conception of compliance in practice. The nature of the problem has to be diagnosed, a remedy prescribed, installed, and made to work efficiently. This obtains even where the remedy may be a simple matter of digging a hole to serve as a settlement pit, for effective action is the central concern. In some cases compliance may involve installation of complex treatment plant requiring heavy expenditure of money and time. Elaborate apparatus often demands regular attention and maintenance to produce an acceptable effluent. Because compliance has an unbounded quality and is subject to constant negotiation, field men do not work to a notion of instant conformity when dealing with persistent problems. Instead compliance in one sense is measurable in months or years, and in the sense that discharges subject to continued monitoring may deteriorate or standards change, compliance is 'for now'. Where states of affairs are concerned compliance strategy requires constant vigilance.

Since water use is a continuing activity it is difficult for a field officer (unlike a policeman: Bittner, 1967b:281) to think in terms of cases being 'closed', which has implications for what he may regard as 'compliance', and for the ways in which he is able for organizational purposes to demonstrate a successful outcome to his enforcement activities. One of the important features of 'compliance' for the pragmatic field man is the existence of some visible evidence of the consequences of his enforcement activity: a cleaner effluent, a new treatment plant, or a discharge diverted from watercourse to foul sewer. Such evidence provides a considerable sense of professional accomplishment:

We drove to a colliery, whose discharges had in the past caused a great deal of trouble, to see two settlement lagoons designed to reduce the level of suspended solids in the effluent. Both had been installed at the field officer's insistence, one at a cost of

£80,000, the other at £60,000. The officer thought the lagoons had greatly improved the quality of the effluent and there was no longer a 'problem'. I'm bloody proud of that,' he said, pointing to the larger of the lagoons. 'I look at those settlement lagoons and I think "That's me what's done that." '

The field officer's evaluation of satisfactory compliance is not geared to legal output measures[5]. So far as the one-off pollution is concerned, he will be satisfied if the discharge is stopped and precautions to prevent repetition are taken. If a one-off is categorized as accidental, the preventive work is often nothing more than some good advice about the need for greater care. A less conciliatory stance is only called for in the rare cases of major spot discharges, which are qualitatively important for their symbolic significance. The pollution which may have occurred as the result of negligence, and especially the persistent failure to comply, require a different approach. Negotiations over the introduction of new or improved pollution control facilities may take months or even years before the officer is satisfied that the discharger has 'complied'. And then compliance must be maintained.

Compliance, then, is much more than conformity, immediate or protracted, to the demands of an enforcement agent. The continuing relationship between officer and polluter, the open-endedness of problems encountered, and the pragmatism of field staff encourage a focus upon the deviant's efforts at compliance, an opportunity denied the deviant in breach of a rule in the traditional criminal code where an act committed is over and done with and beyond repair. A polluter who displays an immediate willingness to take whatever action is necessary may well discover that the gravity of the pollution itself is accorded less importance by the officer: 'it can become a secondary feature,' said one field man, 'if co-operation from the firm is complete.'

Compliance, in short, has a symbolic significance. Enforcement agents need, as much as a concrete accomplishment, some *sign* of compliance. Planning is as important as building; intention as important as action. Assessments of conformity thus tend to be fluid and abstract, rather than concrete and unproblematic. 'Attitudes' are judged as much as activities:

KH How important is the attitude of the other person?
FO Oh, I think that's the most important thing, is his attitude. Because the pollutions themselves can be so variable. ... If he's trying to solve it, I go along with him. If he's not interested in it and thinks 'Well, it will go away in time anyway,' then obviously I'm going to press him harder then. Yeah, it is *the* most single important parameter I think, his attitude. [His emphasis]

The discharger who does what the field man asks—even though he may still be polluting—will be thought of as compliant. Compliance in practice is a continuing effort towards attainment of a goal as much as attaining the goal itself. The extent to which pollution is controlled is no more significant in a compliance strategy than the extent of the polluter's good faith (Goldstein and Ford, 1972:38; Holden, 1966). How 'good' the faith is, however, depends on the kind of polluter encountered.

II. Images of Polluters

It is possible to discern four working categories employed by pollution control staff, of dischargers. Their images are broadly drawn and depict both their individual contacts and the larger organization employing them. These working categories embody impressions of 'typical' kinds of polluter, as well as 'typical' kinds of problem. Polluters are identified by characteristics such as occupation, size, demeanour, and responsiveness. The characterization settled upon is important in contributing to a judgment whether a discharger can be held 'responsible' for causing a pollution (see ch. 8, s.ii).

Most are regarded as *'socially responsible'*.[6] Polluters here comply as a matter of principle, manifesting a 'personal disinclination to act in violation of the law's commands' (Kadish, 1963:437) which spills over into the corporate identity of the firm. Profitable companies and most large undertakings, including the nationalized concerns, form part of this group (see also Katona, 1945:241; Lane, 1954:95; Staw and Szwajkowski, 1975). They are generally viewed as helpful and responsive to the enforcement activities of the agency. The epitome of the 'essentially law-abiding' or 'public-spirited' discharger was said to be one well known company which gives its employees a course on pollution work and its significance, and assigns them precise jobs if a spillage occurs. Though it may impede economic interests, compliance with the law is for some firms a fundamental tenet of company policy. 'Large industries', said an area supervisor, 'have a policy of "We conform with the law—however much it costs".'[7] Similarly, nationalized concerns are expected to be 'socially responsible' because they are particularly concerned about their reputations and also possess the resources to afford whatever remedial action is thought necessary.[8] In a large organization the presence of personalities committed to a policy of compliance in senior positions is regarded as signficant. It is not that the socially responsible do not cause pollution, rather that when pollution occurs it is likely to be defined as an accident, for it will almost certainly be an isolated instance of deviance. The socially responsible discharger in these circumstances will alert the agency, co-operate fully in clearing up, and take steps to prevent a repetition.

Socially responsible dischargers, however, rarely possess unblemished virtue. Most are regarded as reluctantly law-abiding and have to be talked into cleaning up their effluents or taking preventive action. One of the arts of pollution control is to persuade polluters to bear costs in ways which are rarely justifiable commercially, hence the use of the term 'dead money' by field staff to describe expenditure on treatment plant. The image of water users as essentially law-abiding comes with field men's claims of success in achieving compliance in the majority of cases.

Most polluters, however, are half-hearted in their resistance to the demands of the water authority. Those dischargers who cause serious difficulties are in a small minority. This was not always so. The more experienced field staff have discerned a greater pollution consciousness among dischargers and public alike which has been growing since the 1950s. Staff regard this shift as helpful to the task of control. This is not to suggest that dischargers are incapable of deviance, but that serious pollution as a result of negligence or deliberate misconduct is believed now not to be widespread or to occur regularly.

The result at field level is a dual image of dischargers. Staff still expect a measure of resistance to their enforcement efforts.[9] It is portrayed as almost a ritual response to an enforcement agent. For dischargers to 'try it on' or 'try to pull a fast one' is entirely normal behaviour. Disclaimers, evasiveness, and delay are common moves in the enforcement game. Dischargers are expected to 'drag their heels': 'usually people are pretty slow to spend money,' said an area man, 'no matter which sector of the public they come from'. This is true even of those who comply in principle, because delay admits the possibility that one might not have to do what one ought (Silbey and Bittner, n.d.:9). Their hesitancy is not questioned, but is taken for granted to be a desire to avoid commercially unjustifiable costs recoverable only by higher prices. Polluters have nothing to gain from treatment measures: 'It's no skin off his nose if it carries on going downstream. It's one less problem off his mind. He's got less slurry or less whatever to deal with. Or he hasn't got another soakaway to dig.' Many dischargers regard compliance, in the words of an industrialist, as 'doing as little in the way of treatment as they [can] get away with' (Barrett, 1977).

The three remaining categories embrace cases where compliance is less than ideal. First is a group held to be *'unfortunate'*. These dischargers find it difficult to comply completely with the agency's demands, owing to technical inability or to physical or economic incapacity. Agencies recognize that many purification processes are only imperfectly understood and effective treatment is sometimes beyond the competence of available expertise. They also acknowledge that some dischargers do not possess the physical resources (in the form of available land and appropriate topography) to permit effective pollution control—quite apart from the economic costs in terms of capital outlay or potential unemployment which could be the consequence of over-zealous enforcement. Economic incapacity is recognized by field men as a 'genuine' reason for non-compliance, perhaps because for that very reason many of the agencies' own sewage works perform badly. Enforcement in these circumstances produces a feeling of resignation, even of impotence: 'I know before I get there the effluent's going to be bad, and all I'm doing is producing figures so that maybe we can present them at the end of the year. It does seem rather a waste of time in some cases.' Moral inferences as to the willingness of the polluter to comply are not drawn here, in contrast with the two remaining categories.

The third group is comprised of the *'careless'*. Many dischargers find it difficult to adapt to new ways. Many old industries, in existence long before the water authorities or their predecessors, continue to behave as they have always behaved, only to find that the invention of new regulation recasts them as deviants. Then there are other dischargers who, through sloppy management, incompetence, inadequate internal sanctions, or a negligent labour force, regularly fail to maintain their effluents to an acceptable standard, or from time to time cause pollution incidents (cf. Kagan and Scholz, 1979). For field staff the worst form of carelessness amounts to outright negligence, and is particularly exemplified by the discharger whose failure to take preventive measures recommended by them results in pollution. Such negligence is both a symbolic disdain for conformity to the law, and a predictive construct: 'It implies an attitude that will determine whether they will meet our standards and whether I need to be strict with them,' as one officer put it. It is sometimes difficult in practice to distinguish an 'accidental' pollution from a 'careless' one.

Sometimes the images of the polluter which 'explain' the deviance are sharpened in the course of negotiations. Sometimes the image will be redrawn, leading to a shift in 'theory', hence in enforcement practice. Indeed it is possible for a polluter to repair by the extent of his remedial work a previously unfavourable impression, leading to redesignation of the original incident as 'accidental'.

Finally, the *'malicious'* are those who quite deliberately pollute watercourses either to avoid the costs of treatment or disposal, or in symbolic rejection of the agency's authority. Where the careless are ignorant or irresponsible, the malicious are purposive and calculating. They are capable of both isolated and persistent instances of misconduct, but are not now regarded as numerous.

While these characterizations address the reasons for any pollution, another set treat the likely response of a polluter to the process of enforcement. A field man implicitly categorizes polluters into those more able and less able to comply, the effect of which is to shape his stance in the enforcement relationship. Those thought more able to comply are treated with greater stringency and less tolerance of delay or evasion. This categorization is based on a common-sense assessment of the financial well-being of a polluter and a technical judgment about the possiblity of compliance.

A more important categorization, however, one cutting across evaluations of a polluter's ability to pay, and one continually open to redefinition, is a judgment of a polluter's co-operativeness. To regard a discharger as 'co-operative' or having a 'good attitude', or, in contrast, as 'unhelpful' or 'bolshie' informs an officer's expectations about the nature of his relationship with that polluter. Co-operativeness is welcomed for facilitating the job of enforcement and for encouraging principled compliance: 'If you get on well with them, they're more likely to look at the moral issue [of complying] than the economics.' The suggestion of willing compliance from the 'co-operative' polluter announces a respect for the officer's authority and reassures him that his demands are not only reasonably put, but *legitimate.* Besides, a show of compliance is a means of coping with uncertainty, as 'something is being done'.

Characterizations emerge in everyday work and are shared (and thus transmitted to new recruits) in the area office. Informal contact with colleagues creates opportunities to learn what kinds of pollutions and what kinds of polluters 'cause trouble'. The tendency for a label applied to one polluter to be generalized to others defined as falling into the same class is recognized by some: 'You don't trust one of them and it rubs off on the others.'

Past experience is also a source of characterization, favourable or otherwise. The polluter who generates an unfavourable reputation in the agency or who has regularly been in trouble is accorded greater scrutiny and less tolerance when assessments are made: 'You know, if the guy's got a previous history you tend to look on him a bit more harshly than somebody else.'

Officers' characterizations also embody assumptions about a discharger's incentives for complying with the law. It is for this reason that larger companies, with reputations to protect, are expected to be 'co-operative' and 'socially responsible'. Some of the signs emerge from a field man's comments about a large oil company. Responsiveness and willingness to act are important: '[They are] very co-operative because they implement any suggestions we care to make, and they've got a turbo-aerator plant costing £25,000 to improve their effluent. Any suggestions you might make—they'll take note. They'll get things done which they think are necessary.'

But in smaller companies, where margins are narrow and resources meagre, it is assumed that the dictates of commercial rather than legal norms are more likely to be obeyed. Here are found the 'fly-boys' or the 'fly-by-night' polluters who are 'here today, gone tomorrow'. They fall into the 'careless' or 'malicious' categories, whose response will be expected to be 'unco-operative'. These terms are most likely to be employed of small industries in urban areas, especially where 'self-made men' are involved. Such people (it is believed) are motivated solely by profit: in contrast with those who work in large companies who are 'very interested in their image, ... with a little firm they'd just say, "Oh, sod it".' Some rural field men say the same about farmers: 'I find the people with money, that have not been born into it—that are jumped up to it—they're the people that hang on to it and are the most difficult.' The expectation of an 'unco-operative' attitude will be strengthened if the polluter happens to be engaged in a business which produces large quantities of effluent, or effluent which is treatable only at considerable cost, such as the wastes from metal plating.

'Small companies tend to be a problem, the ones that employ ten people, operate on very tight budgets, have no money for effluent treatment, have no technical people to operate what little treatment they may afford. They can be the troublesome ones, I think ... they're the ones that you're constantly going back to and badgering. ... It often is enhanced by a very small site. Effluent treatment and land availability often go hand in hand, so you find that small firms in ... conurbations where they're restricted as to what they can do are the worst polluters I think, without doubt.'

Farmers are an occupational group regarded by officers with rural patches as particularly troublesome, partly because of lack of resources, partly because of a characteristic stubbornness born of decades of water-use unencumbered by the attentions of any regulatory agency. They have a culture of their own which sometimes impedes the officer in doing his job: 'they're a tight-knit community,' said an officer of long standing from a large rural area, 'and they won't inform against one another. There's a lot of cover-up.' Indeed, of all the occupations regularly encountered, the farmer is the most consistently described as difficult to handle, as an area supervisor's question, posed with deliberate ingenuousness, acknowledges: 'Farmers, in particular where you get an unco-operative one, are [some] of the worst customers that we come across. I don't know whether you've heard that before?'

The officer's understanding of the reasons why particular kinds of polluters comply helps shape his choice of enforcement tactics, especially in those cases where the field man expects or is already experiencing 'trouble'. One assumption, with profound implications for enforcement behaviour, is that dischargers are sensitive creatures whose feelings may be easily bruised if urged to do too much, too soon. To 'use the big stick' or 'crack the whip' too zealously may be counter-productive (similarly Kagan, 1980; Stjernquist, 1973). To be too eager or abrasive in enforcement work is to risk encouraging in polluters an unco-operative attitude or even downright hostility. This is a major foundation of the commitment to a conciliatory style of enforcement relying on negotiation as a means of securing compliance; 'co-operation can not be established in the atmosphere of suspicion and distrust that rigid application of the law generates.' (Nicholson, 1973:4).[10] In practical terms this assumption supports two related imperatives for field staff aimed at preserving relationships: 'be reasonable' and 'be patient'. Rather than explicitly seeking to secure compliance at the outset by coercion, officers must demonstrate an understanding of the polluter's problems by discussion and negotiation. Enforcement takes time.

Another assumption is premissed on a belief in the efficacy of individual and general deterrence for organizations concerned with their business prospects and public image. Since businesses and other organizations are regarded as rational institutions which act purposively, they are held to be ultimately amenable to law as a form of control. In practice, the working concept of deterrence is perhaps broader than that contemplated by the law, which is presumably founded on a belief that compliance occurs because those tempted to cause a pollution will be deterred by the threat of the legal sanction. During the research period the maximum fine available to the courts for each pollution charge was £100 on summary conviction in a magistrates' court, £200 on conviction on indictment in the Crown Court.[11] These sanctions were universally regarded by staff of all ranks as inadequate, indeed derisory, deterrents for all polluters except the most impecunious of farmers.

This did not, however, lead to an abandonment of general deterrence as a principal stated rationale for prosecution. Deterrence, rather, resides in the threats which precede use of the formal law, or in the informal sanctions which accompany it (ch. 7). Because field staff intuitively perceive one-off pollutions in cases which are not satisfactorily attributable to 'accidental' causes as the outcome of rational, purposive, or negligent behaviour, they assume that dischargers are susceptible to other kinds of deterrence than the criminal sanction. Where a polluter is being less than co-operative and compliance has clear, easily-quantifiable economic costs attached to it, field staff negotiate by generating contexts in which deterrence will take on significant meaning. They draw attention, for example, to some of the difficulties which the discharger will bring upon himself if he is prosecuted. An individual's or an organization's public reputation is displayed as at risk if non-compliance continues. That aspect of the use of formal law believed to be most important for industrialists ('farmers are different kettle of fish') is the publicity associated with court appearance. Few officers suggest that there is a stigma with economic implications reflected in damage to sales and profitability which attaches to a manufacturer found guilty of polluting a watercourse. Instead it is assumed that a company will seek to protect its reputation as a good-in-itself. Some, however, believe a company's motives to comply are linked with a generalized desire to avoid the 'trouble' stemming from public embarrassment. Publicity implies the construction of reputation, and companies do not want to be seen as 'trouble-makers', since the label will encourage others—especially public authorities which may have benefits to confer—to lay blame for other problems at their door. Other industrialists, it is believed, will not do business with those who engage in sharp practices. The publicity of prosecution can cause all sorts of 'trouble', as an experienced area man suggested:

'It's not so much public relations and the consumer, it's the ... people who live in that particular area who kick up trouble about [certain kinds of industry], for example. ... In general, the public attention is focused on them. They start ringing their MP and all of the rest. ... Every firm likes to give an outward impression of responsibility to the public ... I don't think it affects their sales, y'know, unless it gets to the point of national press campaigns. But ... locally they're very aware of it because they know jolly well if they want to expand, if they want to make alterations and all the rest of it, they're going to get public enemies all the time. ... [The authorities] have got to investigate all public complaints. ... They know jolly well that if they don't satisfy the public, he's going to bring in his local councillor and his MP. [With some firms] you hint about publicity and, boom, that'd be done straight away, that. They're very conscious of it, it's a very useful weapon.'[12]

Another kind of 'trouble' may also encourage management to comply. Internal mechanisms of control may well be an effective constraint against carelessness or misconduct in large organizations, both for senior staff and those at junior levels. 'There's a tendency for any adverse publicity to reflect on the management personally,' said one officer, who had worked in industry before joining the water authority. In other words, individuals likely to be held 'responsible' for a company's prosecution have a substantial incentive to protect their personal reputations and positions (Dickens, 1974). The favourable characterization of the large organization is again supported, as a young officer explained: 'Big companies will jump on their employees. We went round to a big American firm and the Manager said "If we don't get that interceptor within the week, I'll get the sack". So he jumped.' Although talk of the sack in this case may be an exaggeration designed to impress the officer of the urgency with which the firm was complying with his request, the sense of an internal sanctioning system to be taken seriously is clearly conveyed. Indeed, in many firms shopfloor workers held responsible for a pollution which leads to prosecution risk the sack.[13]

The important feature in all of this is the threat of public stigma associated with prosecution for pollution. It is believed to be a more powerful incentive to compliance in more suburban and rural areas where greater value attaches to reputation, and where adverse publicity is more readily transmitted, when, in the familiar phrase, 'the local paper will go to town.' Companies will go to considerable lengths, it is thought, to keep their names out of the papers. The concern about reputation can be exploited by ensuring that maximum publicity is generated about prosecutions, a supervisor explained, to assist enforcement and control in other troublesome cases:

'What you would do is you'd make certain the newspapers were there—being a bit naughty. I've done this time and time again, y'know, a little tip-off to the local newspapers to go along. That used to help a lot. There used to be a splash there. You know, it helps your cause.'

The real value of the stigma, however, is more symbolic than concrete. Prosecution in a compliance system is valuable not as a sanction, but as a threat, because (as the next chapter shows) the threatened sanctions can be made to appear more serious than they are.

III. The Enforcement Game

A major assumption shaping an officer's enforcement strategy and moulding his relationships with dischargers is that most polluters will ultimately do what he wants them to do.[14] Yet most polluters are expected to display reluctance and disingenuousness at the outset. Neutralization (Sykes and Matza, 1957) by way of disclaimers and evasiveness is regarded as a normal response to pollution control work, especially early in an enforcement relationship. The enforcement of regulation, in short, is conceived of in ritual terms. The metaphor of the game is frequently invoked for descriptive purposes (cf. Edelman, 1964:44ff; Ross, 1970:22). 'Everyone tries it on, at least to begin with. Even the big firms who will do what we want eventually,' said a field man from a mixed catchment. 'They will all say "It wasn't us", or "It was an accident".' Asked how often dischargers tried to pull the wool over his eyes, he replied:

'Very regularly, very regularly. Nearly everyone to more or less a degree will try to kid you about something. Either the nature of the cause of the pollution, or how long it's been going on, or they "weren't aware that there was a pipe there", or "Is it really? I've never been down and looked at that watercourse for the last 20 years, I didn't realize we were causing a problem." But nearly everyone tries some minor deception. ... They will all have a go ... even the biggest companies where you're going to get the perfect response, but they will still try to kid you that they "weren't aware that this was happening", or "it was while they were on leave" or, y'know, they "weren't doing it at all".'

Unless, however, the discharger is defined as belonging to a class considered more prone to pollute, or has otherwise displayed himself as unco-operative, his ritual reluctance will be expected to yield ultimately to compliance.

Compliance is not usually achieved without a struggle. Most enforcement problems are caused by a small number of malicious polluters and another larger group of unco-operative dischargers who continue to conduct their affairs as they wish, despite the attentions of pollution control staff. When efforts are made to have them stop or clean up their effluents they adopt protective strategies akin to those which delinquents employ to affect the outcome of their cases (Emerson, 1969:101 ff.). These 'bolshie types' are few in number but occupy disproportionate amounts of field officers' time.

Bolshie polluters employ a variety of strategies to resist, delay, or avoid enforcement. Most officers encounter dischargers who seek to evade detection, or if discovered, attempt to deflect their attentions by resort to various forms of deceit. It is all part of the game. Some polluters go to considerable lengths to pull a fast one: 'They have a man on the look out,' said one officer from a highly industrialized area, 'and as soon as you appear on the horizon, he's off. Or they dump the stuff on a Sunday evening because they know we don't work on a Sunday evening.' A colleague from a mixed catchment gave another kind of example:

'I've had a certain amount ... of y'know making discharges when they think you're not about. I had a farmer who kept me talking and feeding me with tea and biscuits while the farm labourers were emptying the silage tank that had been overflowing into the brook. And then [he] took me along to show me this tank and [said] "Well, you see it couldn't be me, it's empty"—and it was still dripping wet!'

Perhaps the tactic most familiar to field staff is for their admission to premises to be delayed so that incriminating evidence can be disposed of. 'The standard thing', said a supervisor, '... was ... when the [officer] arrived, to keep him hanging about while they turned on the fire mains and all sorts of things so they diluted the effluent. So when he took a sample it was perfectly alright.'

A commoner source of difficulty for the enforcement officer, however, is to persuade dischargers to overcome their reluctance to take effective remedial action, once their pollutions have come to light. Delay is the commonest ploy adopted here. Delay is routinely expected, but difficult to establish, especially where technical questions are concerned, or the polluter has to rely on others for the manufacture or supply of equipment to enable him to comply. Yet an officer cannot afford to overplay his hand; efforts to speed things up may prove to be counter-productive:

'There are always delaying tactics. ... We may be talking here about a method of treating a difficult waste, so the question is how on earth do you treat it? ... Now this can go on for ages and ages [but] they just keep on saying "Oh well, we've written to these manufacturers and the experiments have been going on, but it broke down." ... But you can't prove they're not getting on very quick with it. ... People are very wily and very clever and if they think they're over pushed there is a chance of ... well, slightly delaying and procrastinating on a thing. They always come up with a story of "Well, we haven't got planning permission yet".'

A supervising officer described another kind of delay:

'You come across the technique of never really being able to ... get a decision because you're always chasing the elusive man who can make the decision. ... Some of [the local authorities] are much more adept at playing this game than an industry is. You get a planning committee going. They always *do* something; they're always doing things—working parties, consultants, new committees. And you go round and round and round, and in the end you just have to turn round and say "Look you've got to do something now—or else".' [His emphasis.]

And so far as industry is concerned:

'You've always got works and things going on and you're always digging the place up ... they always employ consultants and they're always looking into problems, they have lots and lots of meetings and things and you can never really screw them down to a day when something is actually going to be finished. ... Some of the big firms do use [the] technique of hiding the person responsible. ... Once you've got as far as you possibly can go with one person and it's obvious that you're getting a bit shirty ... then someone else will appear in the chain of command and you go and see them. And it is a job to find who was really responsible. You know, they ... play this ... managerial game. ...'

In some cases the use of delaying tactics is bolstered by the polluter's attempt to display himself as the blameless victim of *force majeure:* 'They have all sorts of excuses: delay in arrival of spares and breakdowns and this and that; the factory being on holiday and not being able to produce the goods. There's stacks of that; I think that's all part of the game really.'

The use of excuses to deflect accusations of blame suggests some familiarity on the part of dischargers with field staff's conception of the rules of the game. Extensive dealings with pollution control officers make available to polluters a sense of the features of any problem which officers define as significant. The more frequent the involvement with pollution control staff, the more the polluter acquires the awareness of the 'repeat player': he learns the ropes and the procedures, and develops counter-strategies (Galanter, 1974). For instance, the officer who receives a report of a pollution from the polluter himself will be much more favourably disposed than if the pollution came to light as a result of complaints from third parties or even following routine monitoring. The polluter accordingly learns that his own early warning of the pollution is highly valued, from the requests field staff put routinely to dischargers to report any trouble to the agency immediately (Brittan, forthcoming). By alerting the agency when the pollution (however caused) occurs he has a good chance of avoiding the possibility that blame will be attributed to him: favourable moral evaluations lead to more tolerance on the part of field men and a greater reluctance to sanction misconduct (ch. 8, s.ii).[15] Someone who does not alert the agency of a noticeable pollution will be treated prima facie as 'trying to pull a fast one'. On the other hand, a polluter who presents himself as well intentioned may well be able to slow the pace at which he is brought into compliance (but if he subsequently finds himself in trouble, he may discover that his earlier reluctance is now characterized as evidence of a basic unco-operativeness, prompting a less sympathetic approach).

There are, of course, other games, of greater or lesser sophistication:

'You get the attempt to use muscle on a personal level, and you get attempts to use muscle on a political level. You get both. And it works. One sees it working where there is clearly pressure being brought to bear, up in the dizzy heights, and as a result you're restricted in the things that you can do with ... either individuals or companies. Similarly one gets intimidation on a personal level, more so from our agricultural friends than from industrialists.'

In the case which is partly described in Chapter 3:

the officer had received a report of a pollution of a small stream in a public park in the city. The water was cloudy and foul-smelling. The pollution was traced to a small ice-cream manufacturing company, where the manager claimed that a vat had accidentally overturned, spilling the contents down a drain. The officer found the man extremely hostile and aggressive and for this reason decided not to take a formal sample, though in his view there was every reason to do so. The officer thought the manager was lying: the vat had been deliberately overturned, because the ice cream was substandard and tipping it down the drain was the most convenient way of getting rid of it.

Discussing this case later, the officer suggested the firm had probably escaped prosecution because of the manager's belligerence, which made him think first of his personal safety.

FO It wasn't that we didn't want to take a stat; we were anxious to do so, but because this gentleman had been throwing [his product] at us and threatening other forms of physical violence, it was just impossible to get onto his premises to take a statutory sample. In cases like that, one is advised by the authority not to get a broken nose but to retire gracefully and seek a magistrates' warrant, and return. Which is fine if it's a long-term problem, but in the case of a one-off then the intimidation works, and by the time you return then the problem has passed.
KH ... If you are sufficiently belligerent following a one-off pollution then it's very difficult for the officer to enforce the law in effect?
FO Yeah, yeah, ... it can be made virtually impossible if you have to enter land to get the sample.

The bribe is doubtless the best known means of evading law enforcement, though as a taboo subject it is difficult to research (cf. Blau, 1963:187-93). Subtly employed, it can be a useful delaying tactic: 'We used to think that a good lunch would put if off for another three months,' said an officer who had worked, before joining his authority, in an industry notorious for its pollution problems. Sometimes, it seems, polluters are less discreet and attempt to persuade field officers to accept a gift in return for favourable treatment:

'I was once taken to a showroom of a very reputable company when dealing with a pollution. I'd just taken legal samples, and my biggest problem was to get out of the showroom and get on with dealing with the pollution, when he said to me, "Is there anything you want?" Y'know he didn't say "Is there anything you want to drop the prosecution or throw away the samples?", or——he just said, "Is there——if you want anything, just say". And he put it as simple as that. And the cheapest thing ... they made was about £400. ... He made it perfectly clear that, y'know, I could have had what I wanted.'

IV. Bargaining

The voluntary compliance of the regulated is regarded by the agencies as the most desirable means of meeting water quality standards. For the agencies it is not only viewed as the most effective strategy, it is a relatively cheap method of achieving conformity.[16] For agency staff it is a means of promoting goodwill, a matter of profound importance in open-ended enforcement relationships which must be maintained in the future. Compliance takes on the appearance of voluntariness by the use of *bargaining*.[17] Bargaining processes have 'a graduated and accommodative character' (Eisenberg, 1976:654, italics omitted) which draw their efficacy from the ostensibly voluntary commitment of the parties. The more legalistic style of penal enforcement with decision-making by adjudication and the imposition of a sanction risks, according to agency staff, continued intransigence from the guilty polluter. Bargaining is central to enforcement in compliance systems; control is buttressed for it is derived from some sort of consensus (Schuck, 1979:31). Bargaining implies the acquiescence of the regulated, however grudging. And it inevitably suggests some compromise from the rigours of penal enforcement.

The essence of a compliance strategy is the exchange relationship (Blau, 1963:137 ff.), a subtle reminder of the mutual dependence which Edelman (1964:47) regards as central to the conception of the game. The polluter has goodwill, co-operation and, most important, conformity to the law to offer.The enforcement agent may offer in return two important commodities: forbearance and advice.

The offer of forbearance is the opportunity for another display of the officer's craft. He will not ask for costly remedies unless the problem is a major one or the polluter is undoubtedly wealthy. He will recognize inherent constraints facing the polluter, such as lack of space. He will respect a previously co-operative relationship. Most important, he will offer a less authoritarian response than that legally mandated. He offers the polluter time to attain compliance, for bargaining strategies 'are based on the principle that success in pollution control is "bought" by giving up some of the demands that are fixed in the legal norms to be implemented' (Hucke, 1978:18).

Bargaining is possible, then, only *because the law need not be formally enforced.* Rules are a valuable resource for enforcement agents since, as Gouldner has observed (1954:174), they represent something which may be given up, as well as given use. The display of forbearance is valuable in obliging the polluter to take action in response to the show of leniency:

> '... instead of leaving the impression that you're some jumped-up little upstart from an office using the law to tell him what he must do, if you talk to him right, you finish up leaving him with the view that "Well, he's a damn good chap. ... I could've been prosecuted for this. I'm breaking the law, but he's obviously going to shoot it under the carpet and let me get away with it." So ... he does what he has to do, with goodwill, and everybody's happy.'

Or, again:

> 'I've said "Alright, well, look, I've got this stat sample here and this could be used against you in legal proceedings, but provided that you play ball with me and get done what I want to get done, then I'll get rid of it." And [I've] very often poured it out in front of his eyes. And you'd be amazed. This has done the trick on every occasion that I've ever dealt with. They're *so grateful. ...'* [His emphasis.]

Discreetly coercive bargaining is equally useful in preventive work:

> 'You can say "If you build this bund wall[18] or you take these steps, we'll drop the stat." You can achieve a lot with this—*they* feel they've got you off their back, and *we've* got them to do something.' [His emphasis.]

A sense of mutual trust is important in sustaining the bargaining relationship: trust that the polluter will not 'pull a fast one', trust that the officer will not penalize theoretically illegal conduct. Field staff generate a sense of trust by showing how 'reasonable'—that is forbearing—they are, polluters by displaying a willingness to conform and a readiness to report 'problems'. Forbearance aids the detection of pollution by encouraging self-reporting whenever there is an escape of effluent. The polluter himself is, after all, in the best position to discover and control the pollution and prevent it from becoming a public matter. 'You do tend to learn an awful lot more,' said a supervisor based in a major conurbation, 'particularly if they know what your actions are going to be. And also if they realize that when things do go wrong, if they tell you ... you're going to react in a sensible sort of way. And when something goes wrong that they could be prosecuted on, they would still tell you because they know that ... you're going to be more reasonable with them, because you know that they haven't hidden things in the past.'

The field man will do all of this in recognition of the belief that without forbearance, compliance is the more difficult to attain. What he will demand in return for his forbearance is some show on the polluter's part of compliance with his requests:

FO I always explain to them ... that they can do it in active co-operation with us, with us showing forbearance, and we give them as much advice and help as possible. Or they can be awkward and they can do it having been prosecuted ... and with us watching them like a hawk forever afterwards. Which option would they choose? And when I sell it to them that way which way would anyone choose? They are obviously going to opt for the easier option. Now they might try back-sliding, but they can procrastinate once, but then we put dates on them. And legal samples are taken.
KH When you say you put dates on them——?
FO Dates for them to meet the standards. If they say 'Yes, we'll gladly spend the money required' and time goes by and they keep coming up with excuses and it is *clear that they are not in earnest*, then in that case you put a date by which they are required to meet the consent. And if they do not then I feel no compulsion in taking a legal sample. After all, I have tried my best to get them to do it the easy way. And with good feeling all round. [My emphasis.]

The symbolic properties of compliance are clear in these remarks: what is important is the display of good faith.

Since compliance enforcement is practised in a continuing relationship, 'fixing dates' by which a certain degree of compliance must be displayed provides a method of assessing the polluter's degree of conformity. If a period of time is fixed by the field officer (or, better, agreed upon with the polluter) and articulated in the form of a calendar date, the polluter is presented with an unambiguous target by which he should have stopped the discharge, dug a hole, installed equipment or replaced existing plant. Time is a good, exact index of compliance, more useful for enforcement purposes than the prescription of work to be done which is inherently open to negotiation. Deadlines can nevertheless be made flexible to provide the appearance of a softening of demands. Since compliance is often a lengthy and costly process, the field man may spin out the time allowed and present his demands in stages (cf. Kaufman, 1960:225; Thompson, 1950).

Deadlines aid an enforcement officer's appreciation of the polluter's career as deviant. A deadline met is an index of progress, of headway, while a polluter who fails to meet a deadline can be mutually recognized as a rule-breaker, even as unco-operative. In presenting such unambiguous evidence of non-compliance the field man inevitably casts most polluters into a defensive posture, forcing them to account for their failure to live up to their side of the bargain. If the field man can convincingly portray himself as having been generous in the amount of time he has allowed, he can increase the polluter's sense of obligation to comply with future demands. At the same time, failure to meet a date is a breach of the implied bargain and gives the officer grounds to be less forbearing towards the polluter, if this is tactically desirable, unless the latter can offer some plausible account for his failure. What constitutes a plausible account usually involves establishing financial or technical impediments to compliance; thus, to quote an example frequently treated as plausible by field staff, if the polluter can claim the late delivery of equipment he is usually freed from imputations of fault, and has an 'excuse' demanding forbearance from the officer.

The realization that in pollution matters conformity with the law involves more than simply refraining from proscribed conduct is reflected in the second commodity a field man can offer in bargaining. An officer's willingness to act as expert consultant recognizes the often costly and complicated business of compliance. Technical advice will be given whether the field man is dealing with persistent polluters, where the advice will be about the suitability of particular remedial measures, or with one-off cases, where he will want to prevent a recurrence. Advice is always given tentatively, since the officer must protect himself and his agency against any repercussions which may follow heavy expenditure to little effect. Field agents' preoccupation with ends ensures that advice is realistic and negotiable. The remedial action suggested will not simply depend on technical issues centred around the nature of the 'problem', but will also be geared in particular to the officer's perception of the abilities of the polluter to pay. Industrialists are considered to be economically capable of supporting more elaborate measures than other dischargers: 'I think the expensive jobs are loaded on industry and I think industry can bear the cost.' By 'industry' is normally meant companies with large numbers of employees; minor concerns are regarded as 'small men' and *ipso facto* assumed less capable of bearing the costs of control. This categorization between rich and poor is an important distinction, for it provides the officer with a means of establishing whether the troublesome discharger has reasonable grounds for dragging his heels. As most dischargers are not large private companies or nationalized industries, field staff are accustomed to offering advice about remedial action which is as cheap to effect as possible. The more modest the remedy, the more reasonable the demands seem, and the easier the task of enforcement: 'With those that can't afford it I'll say, "Look, just dig a bloody hole in the ground—that'll serve the purpose." I don't like people's money being used needlessly.' And the cheaper the action proposed, the less embarrassing it will be for the officer if his suggestion proves less than successful. The following conversation nicely illustrates the officer's stance, focused as it is on getting the job done 'for now':

FO If there is a problem, a problematical discharge at the site and there are a variety of solutions, one invariably tends to lean towards the least costly, providing it does the job ... simply because you want the industrialist to go along with you and if you go for the most expensive one, you're less likely to get a satisfactory response. The pollution's going on longer while he's umming and ahing and perhaps at the end of the day you've got to take a stat and go through all that machinery, whereas perhaps at the first or second meeting you can agree on a much cheaper, but perhaps equally satisfactory solution. So ... the primary consideration is stopping a pollution as quickly as possible, and if by demonstrating that there is a cheap and effective way of doing it as well as putting an expensive plant in, then in the long-term perhaps it's not as desirable, but by getting that in and getting it cleaned up, you've got breathing space to then keep wearing at them and trying to get them to improve that facility. And in the long-term maybe getting the better plant.
KH So you are saying then that to do your job it's important to know the financial state of a firm?
FO Well you need to know two things really. It's very necessary to be, within a few pounds, aware of the cost of treatment facilities, so that you're not asking for the ridiculous, as well as being able to assess whether the company can afford it, so that you know where to pitch. If there's a plant costing £50 and a plant costing £500 and a plant costing £5000, it helps to know those three costs and to know which one the company can afford, so that you get the right response when you say, 'This is necessary to improve your effluent, you really ought to put this in or put that in.' [You tell them to] get the stuff organized and get it in. But again you're thinking of financial constraints and consultants' fees and, without committing the authority, if you can in general terms give an indication of the plant that they need, give them an on-the-spot consultation if you like, for free, and get them moving ... they're more disposed to doing it than if you say, 'Well, if you ring a consultant he'll come and see you in a fortnight and he'll charge you £200 for driving in through your gate,' and this sort of thing. It's all adding to cost. Obviously one doesn't—can't—give clearly defined guidelines as to treatment facilities because ultimately if the treatment facility fails, then you can't go back to them and say 'It's not good enough,' because they'll say, 'Well it was put in subject to your recommendations.'

V. Postscript

Compliance is often treated as if it were an objectively-defined unproblematic state (e.g. Nagel, 1974), rather than a fluid, negotiable matter. Compliance, however, is an elaborate concept, one better seen as a process, rather than a condition. What will be understood as compliance depends upon the nature of the rule-breaking encountered, and upon the resources and responses of the regulated. The capacity to comply is ultimately evaluated in moral terms, and is of utmost importance in shaping enforcement behaviour. A greater degree of control is likely where a discharger is regarded as able to bear the expenditure for compliance; this issue is still a moral one, fundamentally, not one of economics.

Compliance is negotiable and embraces action, time, and symbol. It addresses both standard and process. It may in some cases consist of present conformity. In others, present rule-breaking will be tolerated on an understanding that there will be conformity in future: compliance represents, in other words, some ideal state towards which an enforcement agent works. Since the enforcement of regulation is a continuing process, compliance is often attained by increments. Conformity to this process itself is another facet of compliance. And when a standard is attained, it must be maintained: compliance here is an unbounded, continuing state. It is not simply a matter of the installation of treatment plant, but how well that plant is made to work, and kept working. And an ideal, once reached, may be replaced or transformed by other changes—in consent, in water resource or land use, for example—which demand the achievement of a different ideal (ch. 2, s.iii). Central to all of this is the symbolic aspect of compliance. A recognition of the legitimacy of the demands of an enforcement agent expressed in a willingness to conform in future will be taken as a display of compliance in itself. Here it is possible for a polluter to be thought of as 'compliant' even though he may continue to break the rules about the discharge of polluting effluent.

A strategy of compliance is a means of sustaining the consent of the regulated where there is ambivalence about the enforcement agency's mandate. Enforcement in a compliance system is founded on reciprocity, for conformity is not simply a matter of the threat or the rare application of legal punishment, but rather a matter of bargaining. The familiar discrepancy between full enforcement and actual practice is 'more of a resource than an embarrassment' (Silbey and Bittner, n.d.:5). Compliance strategy is a means of sustaining the consent of the regulated when there is ambivalence about an enforcement agency's legal mandate. The gap between legal word and legal deed is ironically employed as a way to attaining legislative objectives. Put another way, bargaining is not only adjudged a more efficient means to attain the ends of regulation than the formal enforcement of the rules, bargaining is, ultimately, morally compelled.

7. Negotiating Tactics

I. The Array

A major characteristic of compliance enforcement is its serial and incremental nature, practised in the continuing relationships enforcers must maintain with their clientele when seeking to do preventive work or to control states of affairs. While enforcement of regulation in practice might appear to be conducted in a pliable, constantly shifting, and evolving set of relationships, the amalgam of advice, information, bargain, and threat is carefully organized. The pattern is clear. 'There's a logical sequence of decisions,' one officer observed, 'which you tailor to each eventuality.' To know what action to take in a particular set of circumstances, he continued, is fairly easy—you have recourse to the normal way of doing things in this situation in the past. Resort to 'experience' structures routine ways of handling problems.

For the episodic or persistent polluter, and for preventive work in one-off cases, enforcement is a continuous process, a 'serial testing system' (Rock, 1973b: 92). Its distinctive feature is the application of increments, to use Rock's terms, of increasing unpleasantness.[1] If the polluter cannot be persuaded into compliance by appeals to reason or common sense, then he must be coerced into it by the threat (and sometimes the application) of measures with ever more ominous consequences. Threats play an important part in pollution control since in practice there is little by way of direct action that a field man can take to stop a pollution, such as immediately sealing up a discharge pipe.

The serial enforcement of pollution is an adaptive process, adjusted to specific kinds of violation and specific kinds of deviant. This is in recognition of the many technical and economic impediments to compliance and the many causes of pollution, only some of which may be morally blameworthy. The approach to enforcement reflects the complexity of the problem; the technical and economic capacity of the polluter to conform; and his willingness to comply—his 'co-operativeness'. The enforcement process may be compressed in time and in the number of steps employed in the field man's tactical array:

a 'serious' pollution—a massive one-off—may be handled straightaway with the taking of a legal sample. Here the field man's response is one of immediate sanctioning. A major problem demands a clear display of compliance from the polluter, the appearance of unco-operativeness resulting in the prompt application of weighter measures. In routine cases of persistent failures to comply, however, the enforcement process may be extended and elaborated, with the field man's efforts being dragged out over a long period. The less serious the pollution, the more likely the field officer will be to offer time ('be patient') and to demand less ('be reasonable'). Here the increments of enforcement are often small, and the officer's tolerance the greater. But however insignificant or grave the pollution, field men respond to an unsatisfactory show of compliance by imposing ever more onerous demands. A more coercive posture is adopted, less tolerance displayed, less time granted. Threats of prosecution may begin to be made.

The officer has in practice an array of moves at his disposal in seeking compliance. The tactics are employed serially, the sequence moving from a more conciliatory to a more coercive approach, if conciliation fails. In a prototypical persistent failure to comply, the field man will shift from discussion of the polluter's problem and advice about the remedies which may be attempted (tactics regarded as conciliatory) by way of requests and demands, warnings or threats, and the involvement of his supervisor, to drawing a formal sample of the polluting effluent. These moves are all intended to underline the increasing gravity of the polluter's position. A formal sample may be followed by issue of a Notice of Intention to Commence Proceedings and, finally, by prosecution. Prosecution and the courtroom are treated by staff as symbols of legal authority, and with the drawing of a formal sample and the issuance of a Notice of Intention to Commence Proceedings the symbols acquire a particular immediacy. Though the formal law is the ultimate, logical consequence for the failure of a compliance strategy, negotiation and bargaining will continue throughout, even as the shroud of conciliation drops away.

These moves form a natural gamut from a co-operative to an adversarial stance, although an officer will not necessarily follow the whole sequence, but will break into it or switch to a less conciliatory tactic, by-passing others, depending upon his assessment of the 'seriousness' of the pollution, the 'problem' to be solved, the 'progress' being made, or the 'attitude' of the discharger.

Tactics at the conciliatory end of the gamut are characterized by informality of approach, privacy, and low visibility; as the possibilities of recourse to legal action loom larger, enforcement becomes more formal and more visible, in the sense that the field officer will increasingly involve his supervising officer in the negotiations (as soon as letters are written, matters become more conspicuous to the officer's superiors since correspondence is routed through them). The further the officer moves towards the application of more consequential measures, however, the less control he has over the process, either because he is constrained to preserve his credibility or because his superiors gradually enter the picture.

The culmination of all of this—if all else fails—is prosecution, a step regarded as a matter of profound seriousness which is taken only in a handful of cases a year. Theoretically, a further step beyond prosecution would be the use of an injunction by which the agency could prevent a polluter from continuing to discharge. This is viewed as such a grave response that it is not seriously contemplated as a realistic sanction by most water authority staff and has not, it seems, been employed by either authority studied since reorganization.[2] 'It's always been felt it would be a very heavy weapon,' said a senior official. 'The threat of it, of course, is always there, but it's very blunt in its approach. ...' Indeed, to *threaten* an injunction which is never actually used is a very different matter (see p. 152).

These tactics are not without their risks. In making the increasingly unpleasant response to generate apprehensiveness in the polluter about the extent of his power, the field man must be careful when confronted with intransigence not to become irrevocably committed to the application of a more severe measure than he might otherwise have contemplated. The polluter who becomes inured to the process or who remains flatly obstinate may discover that the field man's powers are not quite so fearsome or readily used as he might have been led to believe, presenting the officer with the problem of preserving his credibility as an enforcement agent. On the other hand, field staff believe more coercive measures invite further intransigence if regarded as unjustified by the polluter (also Cranston, 1979:99). Where the intransigent are concerned, however, the field officer must try to depict the process as inexorable. One consequence of the low penalty structures which normally attach to regulatory misconduct, and another distinctive feature of compliance enforcement, is that it is the act of prosecution itself which is portrayed as the terrible sanction for the offender to avoid, not (as in a sanctioning system) the application of legal penalties of some severity, made possible by prosecution. In the course of negotiation both parties often engage in a continual redrawing of the lines; the field man adjusts his tactics in the light of his interpretation of the discharger's behaviour and degree of compliance, while the polluter adapts to changing demands being placed on him by the officer. Enforcement is founded upon such reciprocal interpretations.

Since remedial action will impose costs upon a discharger which almost always exceed the cost of fines, the essential art of the pollution control officer is to persuade farmers and firms to spend money on something which their commercial instincts would reject as unprofitable. In organizing his negotiating tactics, two central issues inform the officer's use of discretion. His membership and reputation in the organization must first be protected. He must have regard to the expectations of his superiors and to the public position of the organization, matters which emphasize the question of gravity. Is the pollution one in which the water authority must be seen to be taking action? If it is—and relatively few pollutions are categorized as such immediately—the officer's response will be to take a formal sample of the discharge which will be admissible as evidence in court should the water authority decide subsequently to prosecute.[3] The great majority of pollutions, however, are not readily noticeable and most tend not to have readily-noticeable consequences.

It is important for the field man to arrive at some judgment as to the gravity of the pollution, for it has clear implications for his negotiating stance. A 'serious' pollution is one which threatens 'trouble' for the agency—endangered water supplies, or adverse publicity. Serious pollutions demand more vigorously remedial action and field staff are much less tolerant of delay and deviance from polluters. Yet the approach remains flexible:

> 'One site you might take a stat on the second offence, having given them the benefit of the doubt, or the opportunity to make the improvements after the first incident; others there might be a string of minor problems before you really start getting unpleasant. It depends to a large extent on the nature of the discharge and the impact on the watercourse and the——what you consider to be the attitude of the company. One can get situations where they, on the surface, appear to be very helpful and prepared to do a lot, but it never materializes. As I say, it's very difficult to have a hard and fast rule, you just have to assess every one as it comes along.'

Unfavourable moral inferences—pollutions which are 'negligent' or 'deliberate'—also suggest seriousness.

The officer's second concern now comes into play. This is a characterization by him as to cause. Is the pollution 'accidental'? If so, and if the officer is satisfied that repetition is unlikely, then his only concern will be to have the discharge stopped and any ill-effects rectified: fish restocked, sludge removed, oil cleaned up. More sinister interpretations about cause encourage a less conciliatory posture. In fact relatively few pollutions are regarded as purely 'accidental'. Though tankers are sometimes involved in road accidents in which their loads are spilt into ditches, and electric pumps stop during a power failure, field staff recognize there is next to nothing in practical terms which can be done here by way of prevention. If, however, an unbunded oil tank leaks, or vandals break into factory premises and turn taps on, switch pumps off, or disconnect pipes, the officer may demand preventive measures: a bund wall or greater security against intruders. 'Accidents' are not expected to happen where preventive measures are possible. Establishing the extent to which the polluter is to blame becomes a major issue, action and tactics pivoting upon moral assessments. A major source of inferences about moral character is the interaction between officer and polluter; in this, pollution control staff behave like other enforcement agents.[4]

The enforcement process in regulation is highly adaptive. Time and tactical array give the field officer wide room for manoeuvre in most cases without rejecting an essentially conciliatory stance. 'I feel you can crack the whip fairly well', a supervising officer said, 'without contemplating prosecution'. That this is possible is a reflection of the kinds of negotiating tactics employed by field staff and the ultimate willingness of most dischargers to comply.

II. The Projection of Authority

Early impressions of the polluter contribute to the degree of authority the field man needs to present. For the principled, socially-responsible dischargers, it is usually enough that the field man is an agent of a statutory body: 'When we go to people a lot of them work on the understanding that we're from a big statutory undertaking and that what we say goes.' For polluters categorized as 'unfortunate', 'careless', or 'malicious', or in the case of serious one-off pollutions, however, a more explicitly authoritarian stance may be necessary. In general, in terms of the important working distinction between the 'co-operative' and 'unco-operative', the less co-operative the polluter, the more authoritarian the enforcement agent's response. Instead of advice, the field man dispenses orders. The initial presentation of himself as an authoritative person begins the sequence of interpretative work in which the enforcement agent engages, not only shaping the demands he makes of the polluter, but providing the raw material for the development of characterizations about the polluter and his conduct which are of crucial significance for assigning blame and establishing credibility (see ch. 8, s.ii).

The field man, as a public official possessed of power to use the law, is a figure of authority. But since the nature of his job demands a strategy of compliance, he is confronted with a practical problem of how he presents himself to polluters as ultimately authoritative (see generally Goffman, 1959; 1970). Policemen on the street are immediately recognizable as the personification of legal authority; regulatory officials, in contrast, possess no such insignia of authority.[5] They are obscure, anonymous. In the context of the field officer's job a uniform does not connote authority but is demeaning to the individual's status: 'They'd see you coming and think you've come to read the bloody gas meter or something.' (See Joseph and Alex, 1971.) Instead, the lack of distinctive appearance, and the anonymity of 'driving a nice car and wearing day clothes' in themselves paradoxically suggest the individual is a person of status to be taken seriously. Yet for the officer, like the policeman, the appearance of authority and decisiveness is essential (cf. Van Maanen, 1974:116 ff.). To project an image of legal authority is not inconsistent with a strategy founded on compliance, but is valuable as a means of establishing a framework within which negotiations can be conducted.

Voluntary compliance is achieved with a careful exploitation of an officer's position of legal authority, however subtly presented. The law confers an ostensible legitimacy upon the officer's intervention. So much the better if he can manufacture the appearance of representing a community view (cf. Stjernquist, 1973:163-4), and trade on the belief that there is prevailing sentiment that excessive pollution is damaging and must be controlled. But even when dealing with those dischargers who are regarded as essentially compliant and co-operative the field man has to display his demands on their time and money as legitimate and reasonable.

Interaction between officer and polluter is a continuous process of interpretation and reinterpretation of conduct and character. Field staff recognize, indeed, that the business of interpretation is mutual, and the extent to which the polluter will be willing to comply depends upon his interpretation of the officer's effective authority, as the following conversation with a field officer suggests:

KH When you have a problem and you want someone to do something about it, how do you decide what's the appropriate strategy to adopt with that person?
FO Well I think the seriousness of the pollution invariably sort of decides your first response. Is it urgent? Does it need instant action? Or fairly instant action? Or is it something of lesser seriousness which can be solved, well, in time, whatever that particular time factor might be? ... That's your first criterion you've got to look after—are you going to dash into the firm and say, 'Look, you've got to stop this straight away,' or do you say to them, 'I see that pipe is running again, have you still not found the source?' And that is the first reaction. And then I think from there *you simply play it according to the response you receive*. If you get a complete blank from them and they say, 'We are very sorry we've had a spillage of such and such, if you'd like to come I'll show you what's happened and what we're doing about it,' your response needn't take any other form than 'Well in future if it happens will you let us know?' [My emphasis.]

A preliminary task in securing compliance, then, is to establish an atmosphere in which successful negotiations are possible, by creating or preserving a favourable attitude on the part of the discharger.

The approach to negotiation generally adopted can be crystallized into three working rules each of which reflects the priority accorded to the field officer's conception of efficiency. The first, and perhaps most characteristic of a compliance strategy, is '*try to understand the polluter's problems*' because, in the words of a rural officer, 'If you want 'em to spend money to do summat, then you gotta be friendly with 'em.' This obtains both with persistent pollutions and with preventive work following one-offs. An understanding of the burdens imposed by pollution control requirements allied with an appreciation of the practical difficulties often encountered in attempting to comply are considered fundamental to effective enforcement. Compliance is rarely attained instantly.

The display of understanding is helped by patience: 'I always impress on [farmers] if I'm trying to get something done that I don't expect it done tomorrow—I appreciate they've got a job to do—unless it's causing very serious pollution.' And understanding is the more effectively fostered if personal relationships with key people can be established, for this too encourages a co-operative disposition on the part of the polluter:

'If you can get a personal relationship going with somebody, this is a great help. ... It's how you deal with people basically, I think. I think if you try to help and if you try to help and understand, you basically achieve more. ... I try to go in as your "friendly counsellor" as opposed to your "Mr Plod the Policeman" attitude. ... It's all a question of gentle coercion and persuasion. In my experience people tend to react to this. ... It's much better than kicking the door down and playing the high-pressure salesman. ... When people find out you're not there to see them in court, rather [you've tried] to understand their problems, things are much more favourably inclined.'

A display of understanding takes on particular tactical importance when the field man has to deal with firms employing a number of people, especially those whose effluents are difficult to treat, since those with day-to-day responsibility for maintaining effluent control are rarely involved in the negotiation of standards in the consent. Having to devote excessive energy to treat a difficult effluent to strict standards, it is thought, does not always predispose employees to ready acceptance of words of advice or caution from an officer. With farmers, rural field men learn early on that it is essential 'to avoid getting their backs up. You get their backs up and you'll get nowhere.' Too ready a reference to the law here may be counter-productive:

'I quite enjoy going to farms. I try and take an interest when I go there, show an interest in the farm, show them—try and show them—I'm not just someone who comes along from an office in his suit who doesn't know anything about farming at all. I may not approach the question directly, I may have a chat about the farm, see what's he's got. ... You go along and be bloody-minded to a farmer, try and lay the law down, go quoting this Act and that Act which he's probably never heard of and doesn't understand, he'll just put his foot down and you won't get anywhere.'

When dealing with senior staff, on the other hand, personal confidence is an important element in the management of impressions. Often the field man's personal appearance can put him at an immediate disadvantage:

'It's not always easy when you've been out tracing an oil pollution. You might be in dirty wellies and your old clothes. You then go into a factory, knock on the door, and the secretary says, "You'd better see the Managing Director", and you go into someone with a plush carpet. ... You've got to try and present yourself as an officer and you're sat there looking like Bill the maintenance man.'

The preoccupation with demonstrating an understanding of the polluter's problems is a reflection of the interdependence between enforcement agent and discharger, which encourages the posture of tolerance and patience. The commitment to efficiency—to getting results while preserving good relationships in the future—is valued, for it will promote the reporting of a problem, a spillage, an accident or sightings of other pollutions, and will encourage polluters to give an officer ready access to their land to do his job.

Control of the relationship is essential for the field man.[6] This means trading on the discharger's pollution consciousness, or, where the polluter threatens to prove more difficult to handle, displaying technical knowledge or revealing a ready understanding of the particular treatment techniques employed. A wide knowledge of the science and technology of pollution control practice is frequently employed as a front by the field man who feels he may not be taken seriously as a negotiator. Field staff are at an advantage in that they will only rarely encounter polluters who possess more knowledge or expertise than they. 'Expert' advice is the more difficult to dispute and offers less opportunity for resistance.[7]

The officer must constantly bear in mind that of the prerequisites for compliance, the polluter's resources are beyond his control. The field man's only impact can be on individual and organizational will. The problem was succinctly put in hypothetical terms by an officer describing the typical works engineer in a factory: 'Pollution control is one per cent of his workload and it's one per cent he could well forget. ... It doesn't add to production and no-one pats him on the back.'

Field staff often find it helpful to crystallize their negotiating position by correspondence. A letter about the polluter's 'problem' will usually be headed with a reference to the relevant legislation, reminding the recipient that the agency is a law enforcement authority and suggesting the legitimacy of the field man's demands. Letters typically report matters which have been 'agreed upon' in the initial negotiations, acting as a means of making the rules and expectations of the officer relatively clear and unambiguous and as a visible criterion against which further responses may be measured. At the same time the letter is written as a covering device, documenting the field officer's part in the negotiations and shielding him from criticism which might otherwise be levelled at him from above.

A second rule, particularly appropriate when dealing with costly or intractable problems, is '*be careful with threats*'. In deciding when it is appropriate to draw attention to his legal authority, the officer must always be sensitive to the context in which he is working. It is a difficult practical problem, for the image of authority must be achieved without sacrificing the chance of a relationship in which fruitful negotiation can be carried on. So far as the potentially co-operative polluter is concerned, it is important to avoid an excessively authoritarian or officious manner when dealing with an offence; an officer must therefore take care in choosing when to emphasize the fact that his job is ultimately one of law enforcement. Too great a willingness to 'crack the whip' or 'use the big stick' may delay compliance and runs counter to the predominant concern with the expeditious attainment of his objectives, his practical conception of efficiency. But an officer cannot afford to err too far on the side of leniency for risk of appearing to be easily manipulable by the polluter (a 'soft touch'). At some point, an officer must decide whether delay in complying is 'reasonable' or whether it signals reluctance to act. If the latter, it becomes appropriate to make threats about the powers he has at his disposal.

Sometimes threats are discreetly veiled; a reference may be made, for example, to the legal context of pollution control, with the implication that prosecution may be the ultimate consequence. Sometimes they are put in less ambiguous fashion:

> Osgood went to an industrial estate where he found a translucent white effluent resembling milk flowing into the stream on the perimeter. He took a sample. Wandering around the estate looking for the cause of the pollution he discovered two men washing a couple of vans down with detergent. A notice nearby in the yard announced that washing vehicles was prohibited. 'Do you know you're breaking the law?' Osgood asked firmly. 'No? Well, technically you're breaking the law.' The men stopped. 'I'll have to tell the boss,' he went on as we walked away 'because they'll be in trouble from him as well as from me. ... I'll take my sample with us so as to be able to say "Look what's coming down your pipe, Mister".' We found the manager's office where Osgood displayed his sample bottle to the manager and explained where the effluent was coming from. 'If it persists,' he concluded, 'we might have to do something about it.' After we came out he added 'There's no point in pushing them too hard because they have been co-operative in the past. We can't go into it too far because the drainage would have to be examined, [and] we couldn't have them going to foul sewer because the [sewage] works is over-loaded.'

In this case either the field officer's explicit reference to the law or his tone of voice and demeanour achieved the desired effect of making the two men stop washing the vans. Even here, however, mention of the law was mitigated by use of the word 'technically' betraying the field man's view that the breach was insubstantial and would not be followed up. When talking to the manager, the officer made no reference at all to explicitly legal action beyond the veiled and entirely ambiguous threat that 'we might have to do something about it': there was little, in fact, he could do. Besides, they had not presented the signs of a 'troublesome' case—they had been 'co-operative'.

If discussion and advice about possible remedies are insufficient to get the polluter to act, the field officer may employ threats which are calculated actually to strengthen his own relationship with the polluter. He can, for example, present himself as an officer of little discretion and little personal authority; he is, rather, a mediator between the polluter and a potent and inexorable law. Or he is an almost helpless agent who, in enforcing the will of his superiors, is 'just doing his job'. The officer, in the cavalier version of this ploy, projects himself as fighting for the polluter against the insensitive and outrageous demands of his agency: 'I give the impression to people that the authority never loses. ... You keep your superiors in the background. ... You give these people the impression of the authority as this entity, and it's only us that's keeping the entity off their backs.' This tactic is intended to produce in the polluter a sense of personal obligation to the field man, an obligation to be repaid with co-operation, as the following extracts from a conversation about prosecution with an officer from a polluted urban area suggest:

FO ... people are so grateful you've let them off. ...
KH You mean they're grateful ... and this ... helps you in doing your job in the future?
FO Yeah, yeah.
KH Makes you appear benevolent?
FO Because if they get off you claim that it was you who got them off.
KH It was you who got them off, right?
FO Y'see you've always got this nameless water authority at [headquarters]. 'Y'know it's not me, it's these—I give you six months but they'd only give you three weeks.' You can use the authority in that way.
KH I see, you use the authority like that—they're not prosecuted, you can represent it—
FO You can take the credit.
KH —as your handiwork.
FO That's right. ... Obviously you claim credit for anything that will assist you in what you're doing.

Where the polluter seeks to evade responsibility, is unwilling to comply, or adopts an unco-operative posture, the field man will be less reluctant to call upon the hidden presence of the law and will project himself as someone possessed of legal authority.

Sometimes, where the setting is appropriate officers resort to bluffs about the reach of the law in opening negotiations. One tactic, for example, is to portray himself as an authoritative agent simply by using legal terminology intended to convey the apparent relevance in law of the polluter's behaviour. So long as the impression can realistically be created that the pollution is a matter of interest to the law and the officer 'might know something', accuracy and relevance are not important:

FO I don't like to go along and say 'Under the Pollution Prevention Rivers Act 1970, section 7, clause C—', I say, 'Look, the law is this: ...' And if they're the sort who likes to——you have to sense or not whether they like to be confronted with chapter, text, and verse. Occasionally they just like to hear it come out a little bit—they'll accept you a little bit more. No good going along completely vague as to what Act you're talking about. And it does very good to know parts of other Acts that are not definitely connected purely to, not just purely the Pollution Acts. And it's very nice to be——this is a one-upmanship, and there's a bit of one-upmanship in this job all the time. You start ... and say, 'Well, I think that might be in the Caravan Acts of 1960.' I'm not quite sure what a Caravan Act is, I think there is a Caravan Act of 1960, you see. But then again you can refer to 'Trade and Health Acts of 1936'—you see, you're widening your scope. You're making yourself look as a widely-versed person on this subject.
KH It adds to your authority when you're negotiating with them, and persuading them not to——?
FO Yes, it stops people trying to pull flankers on you a bit, because you might know something. 'Town and Country Planning Acts', y'know. 'Section 113'—no that's another one isn't it? But, y'know, I used to know probably more than I know now, I've tended to forget, but when these Acts first came out, I mean one did tend to have to pore

over them when they first came out. ... Even 'Water Acts 1945'—they're relevant y'see. There's certain discharges that can be made to watercourses if it's above a nine inch pipe or something, or less than a nine inch pipe for flushing out systems for the old water boards, which I don't have to know now but I really used to want to know, which I could throw at them. So you should know the law and you should be able to throw in a certain number of other legal terminologies as well, like 'law of tort' and '*mens rea*' and something like that. I'm a bit bolshie, too, given the chance. Anyway, because so often a little bit of knowledge takes you a long way, you're acceptable.

In essence, as another officer put it, 'it has to be a veiled threat; if you can get the job done without using the formal side of it, that's the ideal way to operate.'

When the field man must negotiate with polluters whom he expects to be unco-operative at the outset, however, he will not waste time with an unduly delicate approach. 'The authority is something you can't bugger about too much,' said an officer from a highly-polluted urban area where a large proportion of dischargers were regarded as troublesome, '... at least that's the impression we try to create.'

The use of threats, explicit or veiled, requires the field man to communicate precisely what he expects from the polluter. Officers have to be versatile in presenting themselves for they are obliged to negotiate with people at all levels in a company or organization. Talking to shopfloor workers requires a different approach from that which would be necessary to impress someone of managerial status that the officer is a person to be taken seriously, where a sensitivity to the tone of any encounter is essential to judge when erring on the side of deference might be an appropriate front. With farmers or shopfloor workers, in contrast, the officer may even resort to dialect, a local accent, or other kinds of language seemingly appropriate to the setting which does not distance him from his contact. 'If they start swearing,' in the words of an officer of many years' experience, 'I can swear with them.'

The third working rule is '*don't show you hand*', a precept more relevant for those officers who work in urban or industrialized areas where pollutions occur frequently, where dischargers are expected to be less co-operative, and where it is often difficult to detect the source of the pollution. Here the officer may have to use some bluff or what some of them occasionally refer to as 'chicanery'. He must keep his hand close to his chest to avoid the impression of ignorance or impotence. A practical authority, however, can be suggested by presenting a forceful, but vague, impression of wide and ready power. It is the *suggestion* of knowledge and authority which is the basis of a field man's authority to coerce.

Since the mobility and obscure origins of many pollutions often provide the field man with few reliable clues, the rule is most useful in investigative work. When making enquiries 'you don't let on how little you know when trying to find out who's responsible for a pollution—especially in an urban area where you have culverted watercourses and have to lift manhole covers'. Instead the officer picks up whatever information he can which might be relevant and subtly displays it to appear knowledgeable.

'It happens quite regularly in an urban area ... where my presence has prompted a confession from someone. ... Again it's a bit of sharp talking if you like ... you latch onto anyone you see. [You say something to] get a toe in the door, so that when you get to see the works engineer you've got a bit of backbone behind your bluff.'

The important commodity in being able to pull this off is, again, not knowledge so much as the *impression* of knowledge. If successfully carried off, the bluff may confirm the officer's suspicions or even prompt a confession.

III. Enforcement Moves

Negotiation is carried on in various ways reflecting the officer's own style and his personal judgment as to the most fruitful approach to employ with a particular discharger. Negotiations often open with preparatory work by the field man designed to make the discharger receptive to his requests. A common ploy is to appeal to the polluter's sense of social responsibility. The various uses to which the water containing the effluent will be put downstream can be explained (a particularly appropriate tactic if the effluent is discharged into a potable supply river), or a moral burden imposed upon the polluter by describing the treatment which the water has already received upstream to render it suitable for his own use (this usually takes the familiar form 'how would you like it if you had this sort of discharge coming into your intake?'). Reasoning pursued too vigorously takes on the character of exhortation, a commoner response where implications of blameworthiness on the part of the polluter exist.[8] Some officers, if the setting is right, will use overtly emotive language rather than make apparently dispassionate appeals to what they hope is a sense of social responsibility: 'I tell people they're poisoning water,' said one who worked on a major potable supply river. 'I'll say anything like that to create the right atmosphere.'

A tactic frequently adopted to strengthen the field officer's hand is to portray compliance as ultimately inevitable and the enforcement process as inexorably moving to that end, an approach which can be underpinned by an appeal to financial sensibilities: 'Industry is only interested in pounds and pence. If you can base it on pounds and pence you're 75 per cent home. ... You do some simple arithmetic and you say "Why the hell don't you do it now?" You go on to tell them what it will cost in a few years if they don't do it now.'

For those who claim no great negotiating skill the tactic is expressed more by action than words. They seek to attain their ends by patience and persistence, by making a nuisance of themselves to the discharger, which can only be abated by compliance: 'I wear 'em down. I go back again and again so they don't want to see me any more. ... I get them into the position where they're saying to themselves, "Here's that so-and-so from the water authority again," and they put right what's wrong because they're fed up with me hanging around.' This approach is made possible by a right of access to premises to enforce the law denied the police (Stinchcombe, 1963).

A variant of 'wearing them down', is to put the polluter on the defensive by constant monitoring. This repeatedly brings home to the polluter that his activities and his progress are under close surveillance. With a new pollution a sample takes on a special significance as visible evidence of the problem which can be immediately displayed as a preface to discussions about corrective work. Confronting the discharger on the spot with his poor effluent can have a dramatic value, helping impart a sense of urgency which presentation of sample results some weeks later will lack: dirty water is more striking than computer printout. A sample is also a marker for persistent problems, an index of progress towards compliance. A sample of the untreated pollution is taken, as one field man put it, 'so I've got a record. I then have a sample at the start of the problem to show it was bad. And I can always take one later if nothing else has been done.'

It is important that any encounter with a polluter is managed to give a reliable expectation of the officer's future behaviour. Predictability helps preserve credibility as an enforcement agent, as a man from a difficult urban area explained: 'I never make idle threats, if you like. ... I never say to somebody, "If you don't do that I'll take legal samples", and not do it. They know, people in [my patch] know, that if I say "If certain things don't happen, I will take legal samples", I will take them.' Predictability helps the task of enforcement in other respects; it inculcates a sense of trust in the enforcement relationship, for example, assisting discovery and detection. Furthermore, faced with a polluter unwilling to spend money on preventive measures, a concern for predictability leads to another ploy in which field men create grounds for serious action if a pollution should occur. By presenting their request as reasonable, they establish the basis by which inferences of negligence may be drawn if a mishap occurs in the future. For instance, pollution control officers possess, as yet, no legal powers to compel the bunding of oil tanks. Instead they request that bunding be carried out. This request is in effect a threat masquerading as advice since it embodies a warning that an escape of oil from an unbunded tank is grounds for prosecution, and prosecution is what the discharger should expect.

Another move draws on the rivalry between nearby competitors. A rival's performance can be subtly used to suggest that the polluter has no good grounds for pleading poverty or for claiming technical problems. Discreet enquiries among competitors about the current prosperity of a particular industry provides evidence to create an economic context within which to frame demands to be imposed on others. Close association, whether of industrial rivalry or physical proximity, can be exploited by polluters, however, as a constraint upon an officer's demands. 'I think it's a most difficult situation', said a junior officer to whom the problem was familiar, 'if you get someone saying "You want me to do that, yet him down the road, he's not doing it".' The 'man down the road' becomes an effective means of curbing the agencies' demands, founded as the notion is upon a moral desire for even-handedness of treatment which polluters and officers share alike (see Brittan, forthcoming). The dilemma is constantly raised in those areas where the water authorities' sewage treatment works produce a poor effluent, for they recognize that they 'should adopt an even-handed approach as between their own discharges and those of others' (unpubl. agency document). Field men find they must be careful in the standards they demand of nearby dischargers if their own agency is unable to meet similar standards. In these circumstances the polluter who wishes to delay can do so more effectively in the knowledge that the officer lacks the moral authority he enjoys in negotiations with others. A typical example of the problem was a dyeworks whose effluent was described by the officer concerned as 'black and purple' which 'often' led to 'lots of complaints'. But it was difficult to get the man's contact in the factory to co-operate 'because he knows our sewage works is a far worse polluter than he is'. Sometimes the constraint of even-handedness puts an officer in the uneasy position of making demands which he may regard as otherwise morally unwarranted in a particular case on the grounds that other, similar, dischargers have complied.

We visited one of a cluster of sand and gravel works on a tiny site, where the settlement pit was so small and the machinery and buildings so arranged that the pit could not be emptied by mechanical means. The manager claimed (presumably rhetorically) to have been up to his neck in slurry in the course of cleaning the pit out by hand. Lack of space made redesign of the pit impossible. 'I wouldn't like to have it on my conscience that that chap's up to here everyday', said Lawton, responding to the rhetoric and pointing to his neck, 'to clean out his settlement pit when it's such a small discharge ... it's not important. But we've done all the others, so we've got to do him.'

Where contacts are well established, a series of unspoken understandings about the rules of the enforcement game exist between field officers and dischargers which endow relationships with a substantial measure of equilibrium (cf. Pepinsky, 1976:60-1; Schuck, 1972). This assists the business of securing compliance. But where the relationship is relatively new and there are uncertainties about the rules of the game, or where an unco-operative polluter forces the field man's hand, the officer is faced with a problem of *enforcement credibility*. A polluter who does less than demanded poses a dilemma. On the one hand, the need to demonstrate understanding of the polluter's problems may encourage inactivity: 'If you go too often and talk in a friendly way about their problems they become complacent and don't do anything.' On the other hand, the officer who too readily applies pressure undermines his freedom of action:

SO What you mustn't do, of course, is over-react initially.
KH Why not?
SO If you over-react initially, it's not so easy to escalate.

Enforcement moves are more complicated where organizations are concerned. The problem has implications for deterrence theory and enforcement practice. The theories of individual and general deterrence presumably embodied in the law are premissed on a view of the company as a corporate entity with its own susceptibility to the legal threat. But in the everyday world of enforcement discretion the complexity of organizational arrangements suggests that this is a rather insecure notion (see Reiss, 1980). Enforcement agents only normally deal with a single, responsible individual in very small firms and businesses. Firms are usually comprised of a number of people, many of whom, as individuals, are capable of causing pollution for which the company will be held responsible. A 'discharger' or a 'polluter', then, is comprised of multiple individuals in whom the field man will be professionally interested. Where in the organizational hierarchy does he engage in negotiation in an attempt to exert leverage to secure compliance, assuming this is not determined by company policy? Whom does he negotiate with, urge or threaten when he wants the company to act? Indeed, where a company is concerned, who is 'the polluter'? The person in a position to effect remedial action or the person held responsible for the day to day maintenance of pollution control do not often directly cause a pollution, while the man who inadvertently turns taps on, or spills drums of chemicals may have little personal stake in the company and its reputation. Establishing liability and assigning responsibility demand that the officer should discover what happened and which individuals were involved. In a large organization this is no easy matter.

In the larger forms of manufacturing organization an enforcement agent is confronted with a hierarchy stratified effectively into four tiers: shopfloor workers, supervisors, managers, directors. When dealing with organizations, field officers must cultivate relationships with a variety of individuals of different levels of seniority. The key person is the officer's 'contact', who is possessed of some effective authority over pollution control arrangements, and will normally be seen whenever the officer turns up for sampling or inspection work. The contact's identity and status vary: in some companies he may be a manager, in some the works chemist, in others a foreman or engineer, or the person who actually tends the pollution control plant.

In larger organizations the contact may be of little help in preventive or remedial work. The accidents or acts of carelessness which cause pollution are likely to occur on the shopfloor. The field officer's routine contact, however, may be at a senior level with little or no direct administrative connection with the site of the pollution. The first requirement when seeking compliance is to let it be known in the appropriate quarters what the authority wants to be done. In this the officer has to rely on the efficiency of the organization's internal communication and sanctioning systems. Leverage is applied at the point at which the officer is best able to transmit his demands to get things done, which is often not at the top of an organization, but where the greatest effective concern about the problem is and where sufficient organizational accountability exists to ensure that action will be taken. The particular person chosen depends on the officer's knowledge of the organization, 'But the rule of thumb in cases where you don't know the company', said an officer from a highly-industrialized area who is frequently confronted with this problem, 'is to go for middle management.' Reliance on the internal sanctioning system is important because there is often a problem of commitment—not simply the commitment of people on the shop floor, but the commitment of their supervisors: 'Very often you see [the] pollution side was put under somebody who's supposed to be in charge of it—just another duty added to his other duties and, like all people who have another duty added, it's just another something else he's got to bear on his back apart from all the other things.'

Compliance is a more complex concept where organizations are involved. It is more than remedial action performed by some individual. It demands an exercise of corporate will, a collective determination to make expenditures on plant, equipment, or manpower. Senior executives must be prepared to devote resources to pollution control, and manual workers to do the often time-consuming and unpleasant job of maintaining the treatment plant. 'Responsibility' for pollution incidents is correspondingly complex: the person who is legally responsible is probably not administratively responsible, and almost certainly not the individual who did the act which caused the pollution. The threat of the law may still, however, be unveiled as a negotiating tool to concentrate the collective mind:

'You feel that some people who are really in charge of the operation might not really be fully conversant with what it's all about. So ... at some point you've got to think who it is best to instill the seriousness [in], and that things can happen and they——they can end up in court. "You wouldn't like to be dragged into court as a witness would you?" (You know you personally wouldn't like to be. ... You use a little bit of psychology ... on this side of it.) For someone to think "Ah well, if it goes wrong it's the firm, it's the boss who'll have to go to court not me," but if you start to hint to him that he might be a witness. ...'

The larger the organization, the more difficult continued compliance becomes. Isolated pollutions are regarded as inevitable, even from the most co-operative companies. In many firms hundreds of individuals operate machinery which is capable of causing water pollution. Even though most large companies are viewed as compliant, it is axiomatic for field staff that they will inevitably produce occasional pollution simply owing to the scale of their operation:[9] 'With the turnover you have in these big national companies in terms of workforce ... one gets regular situations of ignorance and negligence resulting in minor discharges, but regular ones that niggle quite a lot.' In such firms pollution is often caused by leakages or spillages, which may be discharged into a watercourse independently of the pollution control plant. In these kinds of case the field officer's concern has to be effectively transmitted to the staff who drive fork-lift trucks or dispose of waste by emptying it into manholes in the factory floor without regard for its ultimate destination. Sometimes, however, other kinds of moves are needed.

IV. Extending Control

The field man's pragmatism makes it fair game to employ a variety of tactics to secure compliance. Where a polluter's sluggishness suggests an unwillingness to comply, field staff may resort to subtler forms of control, seeking to attain the objectives of pollution work by involving other agencies with sanctions to impose or benefits to grant, or by invoking other legislation (see Bittner, 1974; Silbey and Bittner, n.d.).

The water authority is part of a network of relationships maintained by a variety of public bodies which have the task of regulating aspects of the behaviour of farmers, as well as companies and other organizations. Often deviance of interest to the water authority is also the concern of another agency which is better placed to effect compliance. Enforcement is enlarged:

'When it comes to things like oil jetties, they have to be surveyed regularly and given a licence by the local fire officer. And you do a joint survey and you often get the fire officer——you point out to him, you know——You might also slightly touch on his fear if you've got a chance of a spillage, he might be interested as well from the fire safety aspect. So you've got the threat of stopping that licence—it's not your licence, but it's the Fire Brigade, but, y'know, you work together. ... And then, of course, there's public health people. If you've got a problem of an area rather than just one industrial site, then you all get together, because you probably all got problems. ... Obviously it's an advantage to have a concerted front.'

Sometimes field staff may be impeded in controlling a discharge by the procedural requirements of the law. Working through another public agency is often a means of circumventing restrictive procedures. In the following case, the officer was unable formally to connect the pollution to its source:

'[There's a company in town where] they clean car components, rusty car components, by submersion in boiling caustic soda. ... The yard itself is concrete but there were obviously cracks in this concrete and this stuff, which is hosed down every day, had seeped through and accumulated beneath the site. Now this stuff contaminated a pool about a hundred yards away, but it was extremely difficult—well, you couldn't see the stuff going in, although it was collecting in pools ... there was no continuity between the pool of liquid and the pool of water. It subsequently was found that there were land drains between the company and the pool, but I, long before that, I simply went along ... well I simply phoned up [the] County Council and got the waste disposal inspector to come down and we had a meeting with the company and he explained to them that the stuff that was in that pool which had come from their company was a toxic waste, and therefore it was deposited without ... outside the law, and therefore they could be made to do something about it. But we had to sort of take the view that, all right, the pool would be polluted, but it was of lesser importance in this case. But by curing one you cured them both.

There is also considerable scope for the application of authority over reluctant dischargers by virtue of the benefits which the water authorities can confer or withold (cf. Landis, 1938:118 ff.). It is in a discharger's interests to enjoy good relations with the authority, as a supervising officer of wide experience pointed out. The conversation had turned to reasons why companies comply:

SO Because they want to have good relations with us, they see it as a great advantage.
KH ... I mean what's in it for them?
SO Well for one thing, we can manoeuvre local authorities and things etc., for them in terms of getting the dirty stuff into the sewers, in supporting them in developments—because we know they're going to do all the right things—and getting water supplies from the rivers. And also if they have got our support and things go wrong, you know, they know that we will look kindly upon them if we know the firm is playing the game. So it is to their advantage ... to support us.
KH And presumably [there's] sort of a negative of that: if they don't co-operate with you, you will——you are able to exert certain kinds of leverage?
SO Yes. Oh yes.

As the last question suggested, other tactics are sometimes employed which constitute hints or warnings that unless the discharger does what is requested, the agency may withdraw or refuse benefits in its gift. It will—subtly—be made clear to an unco-operative discharger that the agencies derive considerable power from their position in the network of regulatory bodies and from the reach of their own authority into other areas of activity. 'You may look on us as being something you've got to comply with in order to carry on your legitimate business as an industrial person,' said a senior officer of many years experience. 'And don't forget, without our good will it is almost impossible to develop or do anything.' Legislation now requires, for example, that a county council consults the water authority if it is proposing to issue a licence to dispose of waste: this was described by a senior official as an 'obvious' opportunity for the water authority to secure the co-operation of a difficult discharger. A more frequently cited instance involves planning permission. 'The water authority's always consulted on planning matters,' said a supervising officer, 'and I assume that if there's a history of pollution the water authority would oppose consenting to planning applications.' A field officer described the process:

'Certainly with planning applications ... if you're fortunate enough to have an application in the pipeline at the time, or you're engaged in negotiations, then it's a very weighty stick to wave. We see all planning applications [involving use of water]. They go to the various sections for their reply: water supply, trade effluent, land drainage, also [pollution control]. The [field officer] will make comments, if necessary. So, if you weren't getting a good effluent from the factory, and they put in a planning application to extend the factory, then obviously you're in a very powerful position to say "I will object to that on the grounds of inadequate pollution equipment or control. We will object." And I often use that one.'

The tactic can achieve striking results with the recalcitrant. A senior officer gave an example of a company which was engaged in reclaiming metals:

'Pollution has resulted—extremely heavy concentrations of metals have been found in the local surface water drainage system and this has been traced back to the factory. ... Two or three months ago we had a planning application come through because they wanted to readjust the premises and alter them and so on. And we said "No, we will not give our"—I suppose you might say—"approval. ... We won't make any sort of favourable recommendations in this case because we don't know any details of what [you're] proposing." And within, I suppose, a week we had a telephone call from the company saying "Please come and discuss our proposal." Now that's the first time we'd got them to move in any way, shape, or form, reasonably quickly. And we got more information out of that meeting as a result of delaying their planning application than we had done over eighteen months of somewhat patient and tedious negotiation.'[10]

A certain unease about the use of such ploys, however, is suggested in the quite elaborate justifications about bluffing which are put forward in the following conversation:

FO The truth is we couldn't in any way try and withhold planning permission for a responsible development on the basis of the fact that the effluent discharged by the firm was unsatisfactory. But that doesn't stop me hinting that it might be so. [Laughs quietly.]
KH What you are saying is in fact you can't do it, but you can use the bluff again? You can give the impression that you——
FO Oh yeah. I am quite prepared to do that. ... I work on the assumption that what I am doing is basically decent, and right, and to get it done quicker I am prepared to bluff a little bit. We all do that, every day. Everybody has to bluff, and I'm sure that they bluff with me as well, when they come the, you know, the 'You are going to close us down' bit. I am sure that most of them are bluffing—not that it is said to us too often—but there are bluffs on those kinds of lines.

Bluffing is permissible because it is a mutually recognized move in the enforcement game.

V. Bluffing

The field officer's perception of the polluter as rational and responsive to deterrence by abating his pollution if sufficient threat is made, opens up another repertoire of tactical opportunities. A significant part of the general control exercised by field staff is derived from polluters' ignorance of legal provisions and the procedures, policies, and practices of the authority coupled with their status as enforcement agents with the power to initiate prosecution. This ignorance may be exploited by field men in their negotiations by employing various forms of bluff about the legal response to rule-breaking. 'A helluva lot of bluff is used,' said one urban field man. 'You use what tools you've got to exert what pressures you can. And if it's bluff it's fine, so long as they don't call your bluff.' Where the unco-operative are concerned, the field man may threaten the use of procedures or sanctions which he does not intend to employ, or which are not his—legally—to use. Bluffing is a tactic resorted to in other forms of enforcement.[11] In pollution control work it is regarded as a necessary, if sometimes unpleasant, part of the game: 'You have to be more than one-faced in this job. ... I try to be as honest as I can but you do have to bluff your way through certain things.'

There is considerable scope for the use of bluffing in pollution control work since the employees concerned are rarely acquainted with the relevant law or the powers of the water authorities. Even in large companies which may have their own lawyers, contact between enforcement agent and pollution control staff is normally at a point lower and relatively distant in the organizational hierarchy.[12] From the field officer's point of view dischargers know little about the law (Brittan, forthcoming). 'They're not quite certain what you can or cannot do to them,' said one field man. Recognizing that this gave scope for the use of bluff, he concluded: 'I suppose it's in our interests not to let on to people what the law is; rather, you tell them about the law and its penalties when it suits you.' A supervising officer, with due regard for delicacy, suggested that the result of this was that 'We tend to have more impact than there is legislative backing for.'

Bluffs tend to be concerned with legal sanctions or procedures, that is, with the penalties available or the risk of prosecution. In using the bluff, the field man's general intention is to create the impression that the discharger is in more trouble than he might actually be: he is risking prosecution, or he is dealing with a person of greater knowledge or power than is the case in fact. In trying to create an image of the awfulness of the criminal sanction, however, or the opprobrium which will follow, a court appearance, field men do not necessarily believe that a criminal prosecution will produce such unhappy consequences for the polluter. It is enough that the criminal sanction is insufficiently used for there to be considerable ignorance among company staff about the nature of the formal process and the implications of prosecution. This permits a socially-manufactured sort of deterrence. In this way the legislative sanction, and such deterrent force as it possesses, are socially-mediated. Sometimes the tactic will be used quite blatantly where the context is right. Sometimes, however, a field man only need hint about the unfortunate publicity which may follow a company's prosecution. In employing this tactic the officer presents a company's public reputation as at stake, exploiting the desire of most commercial organizations for an unsullied public image. The ploy is viewed as offering substantial leverage for securing compliance.

Some field staff regard bluffing about penalty to be an integral part of enforcement practice in a regulatory agency whose legal sanction is considered by them to be totally insufficient to win the respect of polluters. Their success in this respect may be evident in the fact that a small study of dischargers has shown that most thought the fines for pollution were higher in fact than provided in the law (Brittan, forthcoming). In a job where the practical concern for compliance is accorded such a central position, it is believed to be a natural response to circumvent the problem of an impotent law by portraying it, however subtly, as possessed of a much more fearsome penalty. One field man who works on a large, navigable river is, for example, prepared to tell dischargers suspected of having caused an oil pollution that the penalty on conviction is a £50,000 fine. What he does not say is that this maximum may be imposed under another piece of legislation which his agency does not enforce.[13] If, on the other hand, oil is not involved, the officer will not specify the maximum on summary conviction (at the time of fieldwork £100, though under legislation not yet in force it had been raised to £400). 'What do you think he'd say to me,' he asked, 'if I said [the maximum was] £400, where it's costing him eighty quid a load to tanker it away?' Because the penalty is perceived as derisory, if asked directly what the maximum penalty is, the rule, emphatically, is 'Hedge: at all costs'. To preserve the bluff many evade giving a direct answer. This can be plausibly done: the penalty is determined by the court, and its decision cannot be predicted; the penalty depends on the gravity or frequency of the violation; the court may ultimately impose a sentence of imprisonment, or the factory can be closed down. These latter two statements are pure bluffs because none of the field staff involved in the research could recall an instance in which the sanctions had actually been employed. But the bluff will be used if the setting is appropriate, as the follow-up to the motor works pollution described in Chapter 3 suggests:

We went back the following morning and were plied with tea in the company's pollution control office. Everyone seemed relieved that the problem had been solved. Griffith became a little impatient and began dropping hints that he needed to see for himself, just, as he explained later, to keep them on their toes. Back on the shop floor the cutting oil was being sucked out of the drains: 'They've taken 160 gallons out of there so far,' said Griffith. Lock, the company's pollution control officer, and Griffith both agreed that it would be useful to see the engineer responsible for that part of the shop floor in the hope of preventing a recurrence.

In an office nearby Lock told the engineer what had been happening, adding with great emphasis that it mustn't happen again. 'This is red hot', he said, dramatically. The engineer looked at Griffith. 'What's the fine? A thousand pounds?' Griffith did not answer directly but told the engineer instead that the company could be prosecuted and punished for every sample he took: 'You can be prosecuted just as fast as I fill bottles.' Lock, perhaps thinking that this might be insufficiently persuasive, then asked Griffith: 'You can shut us down as well, Kevin can't you?' 'Yes, I can go to the court,' said Griffith, 'get an injunction, then they'll stick a bung in your discharge and I can shut you down with that.'

Having left the factory Griffith explained his tactics. He agreed that he had deliberately given the engineer a misleading impression of the severity of the penalties, adding 'I need to foster this belief': he had done so because the formal penalties were so low. To let the polluter know what the fines were might encourage him to pollute with virtual impunity. 'I wouldn't go out of my way to disillusion anyone who thought the fines were higher,' he said later, 'because it is more of a deterrent. If a polluter *thinks* the fine is a thousand pounds, then you let him believe it.'[14] This was even truer, he went on, when shop-floor workers, as in this case, were paid piece-rate, since with their earnings at stake they would have no incentive to spend the time in taking preventive measures, or to clear up after a pollution. [His emphasis.]

Another form of bluff can be employed in connection with the requirements of legal procedure. An officer may threaten to take a legal sample of a discharge in circumstances, such as a seepage, where it would be very difficult physically to fill a large sample bottle, or to persuade a court that a sample was taken according to proper legal standards. Or the officer can bluff about the standards in a polluter's consent. Both are integral parts of the effort to secure or maintain compliance:

A dyeworks produced one of the more troublesome discharges in Mullins' patch. There had been discussion in the past with the dyeworks about the introduction of a limit in the consent on the amount of colouring matter which the works could discharge. The dyeworks staff were under the impression that a colour standard which they were already required to observe had been incorporated in the consent, but a colour parameter had not, in fact, been added to the consent form. Mullins did not disabuse the staff of this—and had no intention of doing so—and generally behaved as if a colour standard had actually been imposed.

In this case, the officer was able to represent 'the colour problem' as a difficulty which had to be dealt with, a tactic which enhanced his control in two respects. First, in allowing the company to believe a colour standard was part of the consent, more diligent attention to pollution control would have a broad beneficial impact on effluent quality, since improvement of one parameter often leads naturally to the improvement of others. Secondly, and more importantly, the bluff helped cast the dischargers into a defensive position: they were causing a 'problem', therefore obligated to the officer, in return for his forbearance, to do something about correcting it. With a nice regard for the distinction, the officer described his tactic as 'I don't tell lies, but I do select the truth'.

VI. Postscript

Enforcement activity in pollution control is centred upon the handling of individual cases. Rather than engage in the conscious implementation of a broadly-conceived policy, field staff in this compliance system are more concerned with the management of particular problems. This has the potential for fragmented and particularistic enforcement work since each case tends to be treated on its merits, according to its own special problems. Practical policy is expressed in the accumulation of individualized decisions.

Law enforcement is bound up with the management of appearances (Manning, 1977; 1980). The enforcement of pollution control regulations is no exception to this, at individual or organizational levels. Negotiating tactics are organized to display the enforcement process as inexorable, as an unremitting progress, in the absence of compliance, towards an unpleasant end. If officers fail to exploit the moral compunction arising from the norm of conformity to law, negotiation becomes a matter of the manipulation, delicate or otherwise, of threats. There is an irony here, however. Because prosecution is rare and because legally-imposed sanctions are not regarded as costly to the convicted, field men have to use their negotiating skills to persuade polluters that compliance is in their interests. Many achieve this by portraying the legal process as potent and of surpassing authority. A compliance strategy works, however, because it does *not* use the formal processes of law. Prosecution is, as one field man put it, 'your final ace card', and derives its value from being kept under wraps. The working assumption here is an example of what Muir has called 'the paradox of face' (1977:41): coercion is most successful when injury is threatened, not inflicted.

The preference of the pollution control authorities for compliance enforcement, then, is associated with a corresponding reluctance to exploit the enforcement opportunities provided in the formal law. This in turn inculcates in enforcement staff a belief that their legal sanctions are rather frail. (As Chapter 1 shows, the legal penalties are not as insubstantial as sometimes imagined.) Field staff adapt to this belief and to the desire not to jeopardize effective working relationships with the taint of prosecution by extending their control over polluters in a variety of informal ways. In particular, where law enforcement is practised by negotiation in continuing relationships, a socially-constructed form of deterrence becomes significant. That is, the process of enforcement carries within itself the incentives to comply, rather than the likelihood of detection or conviction, or the punishment embodied in the penalty structure and the sanctioning practices of the courts. The use of threats and bluffs forms a link between legal sanction and enforcement behaviour.

Bluffing tactics further support the field man's view of the elaborate processes of enforcing regulation as a game to be played out with the polluter. In serial enforcement the threats which often precede a switch to a more ominous posture are usually viewed as part of the game: 'the whole thing's a series of bluffs,' in the words of a northern field man. 'The problem arises,' he continued, preoccupied with the question of enforcement credibility, 'when you run out of bluffs.'

The impression of enforcement as a game is buttressed by the belief that polluters are often prepared to engage in bluffs of their own, in keeping with the prevailing view of dischargers, however compliant, as willing to try to pull a fast one. Field men, inculcated as they are with a marked sense of suspiciousness, routinely expect polluters to seek to bluff them about their poverty, their willingness to comply and the efforts they have made to conform. This raises the question of how the officer 'knows' a bluff is being used, and how he treats some accounts as 'plausible' or 'reasonable', and others as disreputable or 'trying to get away with it' (see ch. 8, s.iii). Upon such judgments his future relationship with the polluter and his future enforcement activity will be premissed.

The portrait of the regulatory agency which emerges from all of this is not one which is emasculated by an impotent criminal sanction. Indirect forms of control confer on field staff an authority which, for practical purposes, is in most cases ultimately persuasive.

8. Practical Criminal Law

I. Taking a Stat

Use of the formal legal process is consistently viewed as the last resort in pollution control. The field man's conception of the drastic measure of prosecution begins with the legal sample ('the stat'), for it is this, the critical point in an enforcement career, which comprises the solemn collection of incriminating evidence opening the way to the courts.[1]

The stat is the field officer's symbolic show of strength, its use being described as 'the big stick', or 'hammering the polluter'. It is an interesting reflection on the selectivity and infrequency with which the formal sample and prosecution are employed that the measures should prompt such dramatic phrases, given the relatively low sanctions available to the court upon conviction.[2] The phrases aptly suggest that apart from employing the legal sample to initiate prosecution, field staff also adopt it as another enforcement move, a sanction in its own right, when dealing with serious pollutions. Here the stat is a black mark on an organization's record blemishing field staff's future impressions of it.

If the procedure is properly adhered to, the taking of a formal sample is a much more elaborate matter than the routine collection of effluent. The polluting discharge is drawn in the normal way, but there and then must be divided simultaneously into three equal parts (for which a special three-way funnel is used). One part of the sample is analysed by the water authority; the polluter may, if he wishes, choose one part for his own analysis, while the third is available for independent examination. After the three sample bottles have been filled they have to be closed with special lead seals. Agencies prefer the division of the sample (but not necessarily the taking of the sample itself) to be observed by a representative of the polluter. The polluter's sample bottle is then handed over in return for a receipt: the polluted water is signed, sealed, and delivered. In addition to the effluent sample, further samples of the watercourse itself must be drawn at points upstream and downstream of the discharge to permit evaluation of the impact of the pollution.

Formal sampling is theoretically possible in all the usual forms of pollution encountered by field men, though agency policy has it that a degree of gravity is required:

For the guidance of the [field officer] statutory samples will normally be taken in the following circumstances:

Discharges which result, either alone or in combination with other discharges, in the death or distress of fish.

Discharges which have resulted in justified public complaint.

Discharges which have a marked and serious deleterious effect on the receiving stream.

Discharges which endanger public water supplies.

Unsatisfactory discharges which though not falling into the above categories appear to the investigating officer to be deliberate or premeditated.

Discharges which are persistently outside the consent conditions and where formal (written) warnings have been issued. These will normally be planned statutory samples taken on the instruction of the [area supervisor].

Where there is a persistent contravention of any of the aforementioned enactments, a sequence of statutory samples must be taken so that a number of offences may be established. The interval between the statutory samples will depend upon the circumstances of each individual case, but should not be longer than a period of 21 days. [Unpubl. agency document.]

These guidelines, though relatively detailed, rely heavily, nevertheless, on the interpretative work of field officers. In practice, they are compressed by staff into their two familiar kinds of trouble: major one-off pollutions and serious and persistent failures to comply with consent. One officer put it this way:

'If there's a third party involved and if there's a pollution with a drastic effect then I take a stat. If I come across one on my own causing some obvious effects on the stream, again I take a stat. ... Where it's not going to have other than a visual effect on the stream ... I try to start talking, but if talking doesn't work I take stats. ... If there's a third party involved I more than likely take a stat. The logic of this is ... it gets me off the hook.'

Covering is essential where the pollution or its consequences are not only noticeable but hazardous:

KH What would be the kinds of pollution that you would want to take a stat for?
FO Something that is a danger to life and property, wildlife, I mean. Poisonous matter which could affect public drinking supply, or a fishery where these people with clubs or private people have spent hundreds of pounds stocking the river with fish; and also, of course, for cattle that water probably somewhere on the river.

The agency's sensitivity to its public position buttresses the officer's inclinations, insisting that dramatically noticeable pollutions are formally sampled. The very visible pollutions are normally major one-off discharges. Here, the problems of impact and setting allied with the decay or cessation of the flow of evidence make it imperative for the officer to 'get it in the bucket'. His behaviour in this situation takes on the character of penal enforcement. This is not without its difficulties. When taking a legal sample following a 'one-off' pollution in circumstances where a satisfactory relationship exists between the officer and his contact, the field man has the task of trying to maintain an amicable connection while explicitly doing policing work, for he will continue to be a regular visitor to the premises. This suggests the degree to which the ceremony of the sample itself strikes home, giving the polluter the impression that justly or otherwise he is the target of the law's attentions. If the ritual seems to be resented, the sample may have to be displayed as organizationally compelled, the officer presenting himself as without discretion in the matter. But sometimes it is risky: 'It has happened where there's been very good working relationships with companies over a number of years and ... as a result of a statutory sample the shutters have gone up, and since that time visits to the company have been completely different. One doesn't get the same response from the people you see, and they're less than helpful. Indeed there are some field men who are so concerned to protect the relationships they have cultivated that their supervisors have to instruct them to take formal samples.

The decision is regarded as one of the trickiest areas of the field man's discretion, requiring skill in evaluating person and context. 'You judge pollutions on the situation at the time,' said one. 'I do them off the cuff ... I do things on impressions of the time.' 'It is very much a case of being able to read the bloke that you see regularly', said another,

> 'and knowing how he's going to respond. But of course there are situations where one has a very good working relationship with the contact that you have, but the decisions about improvements are made higher up the tree, so in those cases you have to take the chance of prejudicing that relationship to sting the people above. ... It's very much a case of reading the contact and the company over a period of time.'

Where the impact of the sampling procedure on the polluter is difficult to predict, a stat is usually taken: 'The man who taught me used to say,' one officer reminisced,' "If in doubt, stat 'em, youth." ' The evidence is available if it is needed:

> FO The difficulty arises when you come across something that's fresh, not a problem from a company that you have on the books, as it were, but a new discharge or a one-off discharge where you don't know what the reaction is going to be. And if you don't know what the reaction's going to be then you tend to give them a stat.
> KH You sort of err on the side of caution?
> FO That's right.

KH Cover yourself by at least getting the evidence in the bucket?
FO That's right, that's right.

The problems of decision-making about the use of the formal sample are acute, however, when dealing with persistent failures to comply, which account for perhaps half the formal samples taken.[3] Many persistent failures to comply do not possess characteristics allowing for ready categorization as 'serious'. The conditions which constitute persistent non-compliance are tied to the enforcement relationship, and are more problematic. The polluter may have failed to live up to his side of a bargain; he may have dragged his heels too long; he may have continued to produce a very 'poor' effluent; or he may have exhibited an 'unco-operative' attitude. Many of the signs of non-compliance rest essentially upon a negative moral evaluation by the field officer. The pervasiveness of the moral component is apparent from the following typical description of the kinds of acts or events which one field man said would lead him to take a stat. The remarks are, *inter alia,* littered with references to indicators of blameworthiness:

'Gross negligence by a discharger, for one. If a discharge is being made with the discharger's being fully aware of [it] and knowing that the discharge should not be made, then I would have no hesitation in taking a statutory sample. ... If fish have died I think it is true to say I have always taken a statutory sample. No matter if the discharge has been accidental or whatever, if fish mortality has occurred then I will obtain a statutory sample. If a discharger is persistently discharging [an effluent] which is consented and [it] is outside the consent and the company, or whatever, do not make any attempt to improve the quality of the discharge, then that can also lead me to take a formal sample. If the company have had ... many verbal warnings and a couple of letters warning them and they persistently refuse to accept the advice of the Water Authority, then in that case I would take a statutory sample.'

Use of the stat in 'troublesome' or 'problem' cases is the officer's first concrete threat to confront intransigence. The move comes as the culmination of efforts to negotiate compliance, for when matters drag on unsatisfactorily the officer will have to protect his enforcement credibility and show he means business. When this happens, compliance strategy is in effect abandoned for an adversarial stance in keeping with a sanctioning strategy. The officer's role as an enforcer of legal rules is now made explicit. Conciliation is replaced by confrontation. The field officer is visibly transformed by the procedure, the figure of advice and forbearance now becoming a figure of legal authority.

The threat of the stat is one of the key moves in the enforcement game when dealing with the persistently 'troublesome' case. While asserting the field man's credentials as a legal agent it still allows him the opportunity of displaying leniency if this should prove tactically more suitable. Most officers in fact prefer to use the stat as a bluff and avoid legal action wherever possible: 'With some people, you're gonna do them the first chance you get. We've spent a long time talking, and not really getting anywhere. ... [The stat is] just to make him think we're serious—it doesn't matter if we're not, so long as he thinks so. [The purpose of the stat] is to give someone a push.' The point of the stat as 'frightener' is to make a vivid display to the persistent polluter that he is engaged in frankly illegal behaviour, in contrast with the use of the stat in one-off cases where it is morally deserved or organizationally necessary.

Field officers do not embark on a stat as a frightener lightly, for mundane reasons. They hesitate in the face of the time and trouble necessary to satisfy the legal and bureaucratic requirements: 'You don't rush into stats because you need two people to go out there and then you've got the labelling of the bottles, etc.' Besides, a stat is a diversion from more pressing matters: 'A lot of people think that taking stats is a waste of time because you've got to write a lengthy report that can take half a day.' But at some point, the officer's preoccupaton with his enforcement credibility must lead to a legal sample—even though he may still have no intention of recommending a prosecution.

The ceremony of the formal sample is often valuable in these circumstances. The ritual to fulfil the demands of legal procedure is loaded with symbolic significance. For many polluters it is the first real suggestion of the gravity of their position: 'it sounds impressive—"statutory sample" or "official sample"—especially the sealing business. That makes 'em sit up and think!' The length of the formality, the division of the sample in the presence of witnesses, the careful sealing of the bottles and the signing of the forms all add up to an impressive little piece of legal drama which is often enough to accomplish the officer's primary purpose of strengthening his arm in subsequent negotiation. The sense of legal ritual can strike home, even though the polluter may know nothing about formal samples or pollution law: 'Usually people are quite shocked by the serving of a formal sample.'

Sometimes, however, a stat can prompt a further deterioration in the relationship if the polluter regards it as unwarranted. Officers are instructed that it is sometimes the occasion for antagonism: 'It goes without saying that at all times the investigatory officer must maintain a courteous attitude, regardless of any provocation.' (Unpubl. agency document.) The officer confronted with continued non-compliance following a stat now has some particularly difficult enforcement problems. He must always be prepared to push for prosecution, however distasteful the prospect may appear. The rule is 'you've got to mean business', especially as too frequent a use of the stat as frightener dilutes its deterrent impact: 'You have to be careful with a weapon like this. ... Every time you take a legal sample and you don't proceed on it you lose a little bit of credibility. ... It gets known that everyone has one chance, and you don't want that to get around.' One young field officer who had rapidly earned himself a reputation for his enthusiasm for wider use of his authority's legal powers displayed the dilemma with an engaging metaphor.

> 'If a stat's been taken and nothing happens you think you've fired your gun and missed, and you've only got one bullet anyway. ... Then the enemy knows how big your guns are and they're not very big. ... That's the problem with stats. It carries the force of the law but it doesn't get used in nine out of ten cases—no, ninety-nine out of a hundred.'[4]

The officer must account to the polluter for the failure to prosecute and yet preserve the threat. A trickier task occurs when an officer is put in the position of having to account for a stat which has not been prosecuted on the decision of senior staff, even though he himself wished to prosecute. He has to bluff where, ironically, his stat was not a bluff: 'One likes to feel that when one takes a prosecution sample action's going to be taken. I would prefer to go ahead with it rather than rely on bluffing it out at that stage. I don't like having taken a prosecution sample and not then to prosecute.' Supervisors are familiar with these sentiments and must confront the dilemma of backing their field officer or relying on their own judgment in cases in which they feel prosecution unwarranted.

The threat of the stat is the ultimate practical means of coercion in the field officer's hands. It is a crucial tactic in a job enjoying such a high degree of autonomy, for the officer retains control over the case up to the point at which the legal sample is recorded and reported. Once this happens, effective control over the disposal of the case shifts to senior staff in the organization. Control is of importance in an enforcement system where the great majority of formal samples taken do not result in prosecution and much hinges upon the private transactions between field man and polluter, for while the field officer may be able to cover himself by taking a stat and calm any anxieties he may have personally about the merits of a prosecution, it will be at the cost, if a prosecution is mounted, of loss of ultimate control over the disposal of the case, together with the potential threat to his future relationship with the polluter.

The field officer who seeks to preserve as much direct, personal control over case management will typically employ the stat sparingly, restricting its use to those few instances where prosecution is believed to be clearly warranted to break an *impasse*.[5] This is usually where there are clear moral grounds for action: 'When I don't take legal samples, I don't do so because I lose control of the situation. As soon as I take a stat I lose control because reports have to go up to [the agency head]. I generally only take stats when I think people *deserve* to be prosecuted.' [My emphasis.] Some will avoid going quite so far in the interests of retaining control over a case if some new and immediate sign of good intention from the polluter suggests a willingness to conform. The formal sample, if taken, can then be poured away: 'I've gone to people in the past and said "Look, I've got a stat sample, I'll pour it away on the understanding there'll be no more trouble," ' said an experienced officer. 'And there's been no more difficulty. ...' Then, reflecting upon the private nature of the stat as a bargaining weapon, he added: 'The procedure may be frowned on. ... I'm not sure.' But if a bargain cannot be struck, ultimate control over the disposal of the case passes to senior staff.

II. Assessing Blame

It is in decisions about the use of the stat that the interpretative practices of field officers surface in their most explicit form. Interpretative work is, of course, central to all forms of law enforcement, creating meaning and relevance and providing grounds for action. Enforcement is a continual process of characterization in which inferences about the kind of person and event being handled are drawn from a variety of cues. Characterization imposes structure on an image of reality, making views of the world familiar to others and giving rich scope to the moral inference.

Interpretative work in practice is seemingly instinctive and often unexplained. Yet patterns are discernible. Central to decision making in pollution control is a notion of fault which, when established, prompts allocation of blame and calculation of desert. The field man's working conception of fault is informed by culpability inferred from the pollution set in the context of past acts or omissions, and of characterizations of the individuals concerned. Ironically in a legal context emphasizing strict liability, the guiding principle is a layman's sense of *mens rea*, of constructive guilt, a matter of considerable significance in a control system relying upon such sparing use of formal adjudication in court.[6] Corresponding to the notion of constructive guilt is the fact that circumstances defined as 'extenuating' or 'mitigating' may be taken into account to the polluter's benefit (where, ironically, the formal law would recognize none).

Blameworthiness arises with a persistent failure to comply when it takes on the appearance of wilfulness—where a polluter has broken his consent conditions, as a senior man put it, 'in the spirit as well as in the letter'. Questions of blame are thrown into bolder relief, however, when the field man is confronted with a one-off pollution, for he is immediately drawn to investigation of cause. It is profoundly important for the officer to sort out the moral status of the act and actor: to know whether he is dealing with an 'accident' or a matter with more sinister suggestions of 'negligent' or 'deliberate' wrongdoing. Though knowledge of cause has a practical significance dictated by the imperative to control the pollution, cause is laden with moral overtones about responsibility. Evidence about acts and events is selected and interpreted by the canons of rule-of-thumb assessments, out of which may emerge a moral link between event and actor, and some conclusion about the existence or otherwise of blameworthiness—in other words, a working notion of responsibility.

Common-sense theorizing about cause is the outstanding feature in the interpretative process guiding the use of discretion in pollution control work, one transcending status, area, or authority. There exist three moral master categories as to the cause of a pollution (cf. Emerson, 1969; Sudnow, 1965). In effect, pollutions are regarded as caused by 'blameless', 'negligent', or 'deliberate' behaviour, categories which are intimately linked with the working models of polluters discussed in Chapter 6. Such categorization has profound implications for the enforcement agent's future relationship with the discharger and his immediate course of action.

Blameless pollutions consist of those which are seen as 'accidental' and those which are 'inevitable': both species are blameless because they are not reasonably preventable. While an 'accident' could happen to any discharger, it is more likely to occur on the premises of a 'socially-responsible' one. The 'unfortunate' polluters are responsible for 'inevitable' pollutions. 'Negligent' and 'deliberate' pollutions are preventable, the latter suggesting an entirely free choice in the matter, the former a degree of choice (usually a sin of omission). 'Negligent' pollution is caused by 'careless' polluters, 'deliberate' pollution by the 'malicious'.

These moral assessments offer context to a decision maker: a means of understanding the pollution and the polluter which serves to organize an officer's subsequent relationship and to provide him with rational grounds for a course of action. They also contribute raw material for the further embellishment of characterizations of polluters. The unfavourable moral evaluation is of considerable importance in providing grounds for drawing a case into the pollution control system and, especially where negative moral evaluations of polluters are concerned, in the persistence of a case in the system. The moral offensiveness of the breach is normally a more persuasive factor than the seriousness of the resulting pollution. Except in the rare case where notoriety demands that the agency be seen publicly to enforce the law (ch. 10, s.ii), a moral component is *de facto* a prerequisite for an agency to mount a prosecution.

Important components in this judgment are, again, characterizations of the polluter and his occupation. Whether a pollution is caused 'accidentally' or otherwise is established first by reference to the kind of person making the claim. Officers tend to generalize from particular individuals, acts or encounters to the organization as polluter, simplicity and lack of qualification making the image readily transmissible. Indeed, the disreputability of an individual can stain the identity of an organization.

A second component is the polluter's own account of what happened, for field men are heavily dependent upon such evidence. This account is also tested for plausibility against characterizations of the polluter, his occupation and inferences drawn from common-sense generalizations about human behaviour. Any apparently exceptional attitude or behaviour is routinely recorded in case files by field staff, their portrayals taking on the permanence and transmissibility of the written word. Reflections on the quality or maintenance of the treatment plant, and assessments of the care with which polluting materials are stored and transported on the discharger's premises are also treated as signficant.

Thirdly, field men conjecture about the polluter's capacity to control his behaviour. The exercise of will is a central feature here because will signifies an option—the polluter had the choice to do otherwise (Matza, 1964; McHugh, 1970). The importance accorded these matters is also contingent upon constraints on the ability to comply in the form of natural features such as the site of the discharge, the size of the receiving watercourse and the use made of it, or the amount of land available to the polluter for further treatment of his effluent. If a polluter has serious problems which can be readily understood as not of his own making, such as limited land space or a recession in his industry leaving him—presumably—with limited resources, a more favourable disposition is likely: 'I'm not a great believer in banging in straight away and saying "I'm going to do you," ' said a young officer, 'because you might not know what sort of problems he's got.' A guiding principle for the field man, therefore, is 'You can't legislate for something brought about beyond someone's control.'

A central concern here is the ascription of motive. A large and successful company normally regarded as 'co-operative' is more likely to have a spot discharge defined as 'accidental' in the absence of apparent financial motive to pollute: 'There's very much an intuitive "feel". It's never a specific question ["Can the company afford it?"]. From the scale of the company ... the [field man] is able to make a subjective decision.' On the other hand, a motive to pollute seems clearer in the case of those smaller, apparently less prosperous industries with effluent to dispose of; field men here will be less ready to conclude that an 'accident' has taken place. Reality is often compressed into crude moulds. It is more workable that way: 'I think doing it once is probably accidental. But if the same people do it again ... it can't be accidental.'

Another major element in calculating blameworthiness, one to be found in a variety of legal decision settings, is prior record.[7] Pollution incidents are set in a context of the discharger's past behaviour. A pollution is not normally regarded as a discrete event but is seen as part of a career, a continuous line of conduct contributing to the context in which a particular incident may be judged. A past history of deviance suggests that a specific pollution is merely a symptom of a clear propensity for continual breach of regulation; in some cases it may suggest a calculated willingness to flout the law allied with deliberate disdain for the efforts of enforcement agents. Suggestions of 'accident', on the other hand, are easier to accept if there is no history of 'trouble' from the discharger: 'I take into account past record,' said a supervisor,

> 'and [when] somebody comes along with a genuine accident and they've never had any trouble before and they've always complied, then you say to yourself, "Well, they've done their best and this was an accident," and let them off. In exactly the same set of

circumstances, if they came along when they'd had a history of waving two fingers, if you like, at the authority ... then you'd say "No, no way will we let them off," y'know, "they're always doing it." So, ... they are storing up trouble. ...'

In the absence of a prior delinquency, a one-off from a discharger will normally be presumed to be an accident unless there is conspicuous evidence to the contrary. A history of ready compliance makes a pollution incident appear incongruous, transforming it into 'an isolated instance'. The shift has major implications for the officer's use of discretion:

'I know of companies where if I caught them in a pollution I'd have no hesitation in taking a formal sample because they'd been bad boys in the past. But there are others where I'd need evidence of massive negligence before I'd even begin to think of taking a formal sample. Unless, of course, the consequences for the river were very serious indeed.'

Reluctance about sanctioning dischargers who have not been 'bad boys in the past' is clear in the references to 'massive negligence before I'd even begin to think ...', and especially so in light of the fact that this officer's rivers were so dirty that only a huge pollution would produce noticeable consequences, quite apart from 'very serious' consequences. A 'good' past history, then, begets tolerance from enforcement agents judging present deviance. In pollution control work this is expressed in a working rule of 'give a dog one bite', which resembles the practice in juvenile justice of 'three strikes and you're out' (Emerson, 1969). 'I think we have always said "Well, they can have one bite of the cherry or one bite of the apple," ' said an area supervisor,

'providing ... that it's not a particularly bad type of effluent, such as a cyanide-containing effluent, where the firm should have known that the material was particularly hazardous and would kill fish, hazard the population at large. ... The accidental part about it is that one allows that one bite, gives them a warning to put their house in order, and doesn't normally prosecute on that first occasion. This has always been sort of historical with us.'

More recently, he continued, there was evidence that his agency was becoming increasingly sensitive to the problem of discerning the plausibility of a polluter's account, as the following reference to 'so-called accidents' suggests, resulting in some change in the 'one dog, one bite' rule:

'There has been a tightening-up in respect of these accidents or so-called accidents. In fact we've had instructions [from the top] that whereas in the past we might have regarded a first instance as a once-off thing that might not be repeated, and you warned the firm, there are occasions where a once-off pollution should be regarded as either deliberate or carelessness, and a prosecution is warranted.'

Past history is equally relevant when handling persistent failures to comply. Prior record in these circumstances possesses a predictive component, the reasoning suggesting that past and present deviance will continue in the future unless some enforcement action is taken. Prosecution is only partly justified by this utilitarian rationale, however, for a moral justification—the greater the persistence, the greater the desert—is also a powerful influence. The polluter interpreted as consistently failing to accept advice or observe a bargain is more deserving of sanction, hence the response of one officer from a highly polluted urban area who claimed that he would take a formal sample 'for revenge' if 'led a song and dance' by a discharger whose delay in effecting remedial action caused a pollution.

A third source of moral character is the various cues which emerge in the process of control, particularly those attaching to the discharger's demeanour and disposition. An image of a discharger as pleasant, helpful, and co-operative will usually serve to deflect suggestions of blame. For instance, in the following 'blameless' case:

Griffith was standing on a bridge in his urban patch when he noticed a thick white sludge on the stream bed. The sludge was clearly observable from the pavement, its public visibility causing the field man particular concern. He spent the best part of half a day lifting manhole covers and studying maps in an effort to track the sludge to its source. He finally pinned it down to a plating company on an industrial estate nearby. Griffith thought the pollution bad enough to deserve an immediate stat, but then remarked that the case was 'not as simple as that'. The people who owned the estate, he explained, were very nice, very helpful, very co-operative. The discharge, legally, came from their property, not from the plating company which was their tenant, so they were responsible for it in law. He could not take a stat, he said, because the owners of the estate were not to blame for the pollution, even though it was a bad one.

The polluter's demeanour and behaviour in encounters with the field man are components in the ultimate evaluation, underlining Lemert's dictum (1972:59) that legal control is often a response to the style rather than the content of behaviour. Field men are sensitive to the same sorts of cues as policemen and respond in much the same way.[8] Rudeness, disrespect, or hostility invite a less favourable character assessment and encourage a less conciliatory approach (excused on the grounds that 'we're all human aren't we?'). Affronts to the officer's authority are against the rules of the game and call for sanctioning: 'I think you wave the big stick when you feel you are being used. Y'know, they think "Oh, here's a right one, here," y'know, "we'll fob him off." ... So this is where you start to say ' "Right, monkey—that's what you think." ' There is an inevitability about the evaluation:

'I think that by and large if a man is sheer bloody-minded and tells you in other words that you're a lot of b.s., and what the hell are you doing there worrying him, and its your job to clean up the oil ... I would put two and two together and say that his outfit is in every way responsible. And I don't think in this particular case one can avoid the conclusion that it was sheer negligence on the part of the company.'

'To the helpful sort of person, I'd be helpful,' said another officer. 'This is a rule of life.' The enforcement agent relies again on common-sense knowledge and assumptions about human behaviour to decide whether a person is being 'genuinely helpful'. The most commonplace behaviour offers clues to character. 'There are some con merchants,' the officer continued, 'and I've seen some. But you tend to spot that kind of person. He's too nice—he takes you down to the pub and buys you lunch.'

Many opportunities for character assessment reflect the field man's preoccupation with efficiency. The polluter who reports a pollution on his premises is treated sympathetically, partly because he has given the agency notice, and partly because to punish self-reporting is to risk deterring the offender and others from doing the same in the future. But reporting a pollution oneself and helping clear it up are also symbolically important signs of a willing admission of responsibility, an expression of an identity of values shared with the control agency, akin to the signs of remorse in offenders rewarded by criminal justice officials. The admission carries hints that the pollution was not intentionally caused for it suggests that the polluter has nothing to hide. Wrong-doers, it seems, rarely confess. Informing the agency is also a means by which the polluter is able to fulfil part of the implicit bargain between dischargers and agency: for its part the authority will forbear from sanctioning. To punish the polluter after such a symbolic show of regret goes against the grain, quite apart from the risks to future co-operativeness:

'With an accident, I take a lenient view. Depends on the size of the pollution. Accidents happen. [In one case of an oil pollution which was the more serious for being on a potable supply river] the management had taken every reasonable step to avoid pollution. Just some berk hadn't shut down a valve properly ... but through some ignorant action by a shift worker in the middle of the night, this was why the oil came out. It was accidental and perhaps a bit negligent as well. ... They were about to phone us up and let us know. [That was why] I went for a stiff warning. ... I told them if they have a spillage and they let us know [we'll take a reasonable view] because it allows us time to do something about it. I always tell firms if they have problems, give us a call and we can help them out. So if you go round beating the big stick you won't get much co-operation. But if you bend with the wind you get more. ...'

Interpretative work is a precarious thing, however. One seemingly trivial act or event, one nuance in the polluter's demeanour or behaviour, may be enough radically to transform a characterization. To report a pollution which has taken place on the premises will be applauded by the agency. But if done 'once too often' it will be regarded as 'trying it on'. In effect the polluter has to conform to certain expectations held by the officer as to what constitutes 'proper reporting': 'If it is reported and dealt with responsibly then it's a different kettle of fish than if there's loud screams and shouts a couple of days later ... complaints coming from all sorts, turning [field staff] out at night-time, and then the problem being far worse than it originally was.' Like the polluter's other activities, self-reporting behaviour is also put in its historical context:

A field officer was having difficulties with a large manufacturing company which was frequently telephoning him to warn him of pollutions on the premises. This happened so often, he reasoned, that the company must be using it as a tactic to mask its incompetence, having learned it could get away with it by making a phone call. The next time the company warned him of a spillage, his response was to take a formal sample. [From field notes].

A polluter's acts or omissions after a pollution has come to light offer further evidence of his moral worth: 'the fact that it's an accident to start with doesn't necessarily mean that I'm going to look upon the whole thing as an accident. It's how each company deals with it from there.' An unwillingness to help constitutes a display of indifference, persistent denial of responsibility a suggestion of calculation. On the other hand, efforts at repair mitigate culpability: work by the polluter to minimize the impact of the effluent, to clear up the mess, to pay for remedial measures or restock a watercourse with fish symbolize a desire to make amends. These are grounds for attributing less blame than might otherwise be deserved. Pollutions followed by remedial work are more likely to be categorized as 'accidental' or 'careless' rather than 'negligent' or 'deliberate', and lead to a more conciliatory posture than might otherwise be the case: 'If the firm are helpful—you know, open the gates, let your men in, help you clear things up, etc—it does, I mean, it's *forced* to affect your opinion of people.' [His emphasis.]

The blameless pollution is not often followed by anything more than suggestions about preventive measures or warnings as to future conduct. No sanction is exacted unless the pollution is very grave and visible to the public:

A massive pollution occurred in a major potable supply river when an employee at a recently-built plating company flushed out a vat containing chromate solution. The employee turned a tap on under the mistaken impression that the contents of the vat would be discharged into the foul sewer. When the factory was built, however, the contractors had wrongly linked the vat's effluent pipe to a surface water drain connected with the main river. As a result of the pollution many of the potable supply intakes on the river had to be closed down and an emergency water supply arranged for a town nearby. Pollution control staff worked round the clock in continuous monitoring and cleaning up of the slug of chromate solution, the agency spending over £10,000 to protect the public water supply.

The company were prosecuted in the Crown Court, such was the seriousness with which the agency viewed the incident. The defendants pleaded that it was not they who had 'caused' the pollution but the contractors, by wrongly connecting up the effluent pipe. While the court accepted that the company did not know the pipe had been wrongly connected when the tap was turned on, it did, however, find it guilty of causing the pollution, following *Alphacell v. Woodward.*[9] Most agency staff also believed that it was the contractor, not the company, who was to blame. But from the agency's point of view such a grave pollution demanded a striking public response with the unusual and dramatic step of prosecution in Crown Court.[19] [From field notes.]

This case shows that the agency imperative to be seen publicly to be taking action may lead to serious consequences for the polluter, surpassing any moral features which would otherwise have mitigated the common-sense conception of culpability. But these are rare circumstances, and staff normally take a different view. 'It's a bit moral, really,' said one officer, discussing the propriety of prosecuting a discharger for an accidental pollution, 'a lot of us would think it's wrong.' Indeed, this judgment has extremely important practical consequences, for only in the very exceptional circumstances outlined above will the strict liability law be invoked following accidents.

The negligent pollution is a much more problematic matter encompassing varying shades of blameworthiness. Negligent behaviour is a catch-all category embracing a range of conduct from simple carelessness ('bad housekeeping') to culpable disregard of an officer's instructions or warnings. This latter form of negligence is regarded as a weighty matter, and may well be sanctioned by a formal sample and prosecution where a bad pollution has been caused. The essential difference between 'carelessness' and 'negligence' is that in the latter, knowledge or foreseeability, actual or constructive, is present: 'Carelessness might be because somebody had not tightened a pipe up correctly ... but if a pipe had been allowed to corrode over a period of years, then I would regard that as negligence. ... Negligence I think is something that could have been foreseen for a long time in advance.'

The warnings given by field staff as an integral part of negotiation are useful in establishing clear-cut grounds for inferring negligence, as an unpublished agency document points out: 'a repetition of such pollution after due warning and the giving of advice by the Authority on how repetition reasonably can be avoided would be classed as arising from gross neglect or carelessness.' If the field officer is satisfied that the discharger knew or ought to have known that pollution would be the consequence of some act, event, or omission, then grounds for inferring negligence are present. This obtains even if the event actually causing the pollution (a vehicle colliding with an unbunded oil tank, for example) is itself readily interpretable as an accident. The point is that the owner of the tank is held responsible for the oil spillage by ignoring a warning about the dangers of unbunded tanks.

Where no explicit warning has been given, testing for foreknowledge can be done by raising simple questions: has the pollution happened before? Had appropriate precautions been taken? Had there been publicity about this kind of pollution? Publicity is a ground for assuming that the polluter should have known, and that the pollution which subsequently occurs is prima facie due to negligence.

A deliberate pollution is not a common occurrence and is treated as a grave matter. A calculated act of pollution in defiance of the law is a symbolic assault on the field man's river and provokes moral indignation which overrides any niceties about maintenance of relationships by conciliation: 'He wants doing good and proper,' said an officer of one case. 'That's pure malicious pollution: he doesn't give a tinker's for it; he wants doing good and proper.' In practice a pollution with serious consequences which is defined as deliberate is a prime candidate for prosecution.

There appears to be a relationship between the seriousness of the pollution and the readiness with which negative moral evaluations are made of it. The more serious a pollution, the greater the tendency to assume that it could not have occurred by accident. The greater the damage, it seems, the greater the likelihood of staff finding fault and attributing blame.[11] It is as if an officer cannot conceive of a serious pollution coming about 'by accident', perhaps because an 'accidental' cause suggests that the polluter has failed to take proper preventive measures, for which the officer feels to some extent responsible. In the chromate case described earlier, some of the staff involved in monitoring and cleaning up were of the opinion that the company 'must have known' what they were doing when the tap was turned on, despite the court's finding of accident.

Where there is an absence of evidence to link act with intent a *notion of plausibility* is employed to tie the two, permitting the argument that some element or other suggesting negligent or deliberate conduct 'must have been' present. In such cases, if a pollution can be traced to its source, the officer's next step is to construct a plausible account for himself of how and where the pollution took place. In doing so, field men have to rely heavily on accounts offered by polluters in explaining untoward behaviour (Scott and Lyman, 1970).

III. Credibility

Witnesses to polluting activity are hard to find and the cause of many pollutions can be readily concealed. Explanations offered by those who break rules, however, are suspect, and field men must devise some means, however banal, by which the credibility of the polluter may be assessed. Judgments about fault and blame rest on the procedures which field staff employ to establish for themselves the truthfulness of any account put to them by polluters.

The way in which versions of 'the truth' are found is inextricably bound up with moral assessments. For instance, the account of a polluter deemed, on the basis of encounters and past history, to be disreputable in some way is unreliable. And the discovery that a previously 'co-operative' or 'helpful' polluter has 'pulled a fast one' beyond what is routinely expected is important in inviting continuing suspicion from field staff.

Assessments of credibility draw on a variety of evidence. Familiarity with pollution treatment processes equips the officer to perform 'uncovering moves' (Goffman, 1970), piercing proferred explanations, while familiarity with the discharger and his problems offers the possibility of corroboration: 'You've got to go to a place lots of times and get talking to lots of people. And you've got to listen to other people's conversations. ... You've got to know people.' Bluffing is routinely expected of polluters as a part of the game, and widening the enquiry in search of corroborative evidence in conjunction with the officer's experience equips him with a perspective to test for truth:

> 'You always talk to more than one person in any firm. There is always somebody in the firm who will give you the low down. Sometimes they are being very unfair, of course, and you have to watch out for that too. But in the end you have got to make your own judgment. ... There aren't many bluffs that they can use against me that are really of any significance anyway. I mean [their] threat to close the place down in the end wouldn't make any difference, I mean it is a very, very peripheral affair; and while they do try and bluff me—and I have heard some wonderful bluffs in my time—I kid myself that I could see through nearly all of them.'

To test for bluffing calls for an inspection of the evidence. For instance, accounts which recur arouse suspicions:

FO 'Not many people pull the wool over my eyes—well, we all think that, but—after five or six years you—you'd be surprised how perceptive you can be about people in summing them up. You do learn. You make mistakes of course, but you do learn. And you find the regular excuses and bluffs, and you have always got an answer for them.
KH You mean the same old excuse they use—
FO Yeah, of course they are—
KH—and the same old stories, over and over again?
FO Yes, yes.

The internal consistency of any account is another simple test:

FO Most of us are bad liars and if you're going to be a good liar you've got to remember what you've said previously. They start to give it away. Little things they say. You've only to listen to their conversation and they soon realize that—
KH You look for inconsistencies?
FO Yes. In conversation, and in common-sense things. Farmers—if a farmer tells you that he's farmed that farm and his father before him and his grandfather before that, there's no way that he can kid you that he didn't know that that pipe went to that particular brook. He could probably tell you exactly how deep it is and when it was put in. But it's very common for them to say 'Oh I didn't realize there was a pipe taking that effluent to that particular point.' There's no way. If the man's only moved into the farm a couple or three years ago then he's quite likely to be telling the truth and he may well not know where his land drains or whatever they are go. ... But more often than not ... they slip up during conversation, they start to contradict themselves on something or other, whatever deception they're trying to pull. It is a little bit annoying. You feel that they're treating you as less than intelligent.

It is taken for granted as part of the game that polluters will protest their innocence. Too vigorous an effort to do so, however, prompts the suspicion that the polluter is committing a foul by seeking to hide something significant. This is another reflection of the enforcement agent's preoccupation with the incongruous: where neutralization of guilt is routinely expected, vigorous protestations of innocence are suspect.

Again, cues picked up during an encounter with the contact or others may be treated as relevant signs of credibility. Demeanour is particularly important: 'Some might affect cleverness. If a man's in the wrong, very often he'll become aggressive. He'll adopt the attack defence, so to speak. ... If he's aggressive he's got something to hide, almost certainly. If he's clever and smart, he's got something to hide.'

Judgments about credibility are made routinely at every stage in the enforcement process. Field men find it essential, for example, to be able to assess the truth of dischargers' claims about their lack of resources since pleas of poverty are so regularly encountered as excuses for non-compliance. The raw material for an assessment of means may be based simply on the most mundane appearances:

tumbleweed blowing in the yard, said one officer, is evidence of a firm in dire straits. Such visible evidence can be buttressed by knowledge of the local industries and their current profitability; the local newspapers will report if men are being made redundant; the national press will present evidence of a company's profits. Sometimes field men in industrial areas will talk to a company's competitors ('playing one firm off against another'), on the assumption that a discharger will be more honest about a rival than about himself.[12] Often an officer has little to go on in assessing pleas of poverty, especially where he is dealing with farmers and small dischargers whose effluent is less frequently sampled and with whom he will be less familiar. The trivial and the banal are all potentially useful as evidence. 'Sometimes you get invited in for a cup of tea,' said a man from a rural area, 'and you drink it out of a cup and saucer. But I've drunk it standing up in the kitchen out of an iron mug.' With less information it is recognized there is greater margin for error. Appearances can sometimes be deceptive. 'He showed me a slurry tank,' said an area man wryly. 'Turned out to be a swimming pool.'

Searching out the truth is another piece of the pollution control officer's art acquired in the job. Just as the policeman's stereotypes crystallize with experience (Rubinstein, 1973), the competent man can quickly make up his mind:

> 'You've not got to be in the job long to know what's going on. ... Normally when people lie, it's done so badly—so badly thought out—it's obvious. ... I had this case the other day when there was an oil pollution on a building site and the building foreman said, "We were digging a hole and the oil started coming out of the ground." And I just said "Oh yeah. When are you going to tell BP about it?" '

IV. Postscript

Moral judgment is all-pervasive in polution control. For example, a compliance strategy is morally inappropriate in cases of wilful rule-breaking, whether it be in the form of deliberate wrong-doing or persistent unwillingness to comply. A sanctioning strategy is apt because in either case the offender has chosen to place himself outside the enforcement relationship and outside the law, a choice deserving of punishment. Both deliberate rule-breaking and non-compliance with the enforcement agency amount to symbolic rejections of that agency's authority and legitimacy. While it may be instrumental to threaten the use of prosecution in compliance strategy, recourse to prosecution transforms matters into the realm of the symbolic.

Enforcement is also open to transformation from compliance to sanctioning system (or vice versa if new moral data produced in the course of encounters with the polluter allow refinement of the officer's working theory of rule-breaking).

The moral undercurrents informing even the most banal of enforcement work are mutually recognized by regulator and regulated. A sanctioning strategy can be risked with culpable rule-breaking or unco-operativeness because the agency assumes the polluter—and interested others—understand that it has good reason for taking a tough line. The polluter whose rule-breaking occurred by accident, on the other hand, is likely to view such a response as unjust. The conceptions of moral fault operating here are, however, rather wider than the law's conception of *mens rea.* Common-sense conceptions of culpability reach beyond the polluting act into more general realms of disreputability. As they do so, the urge to punish gains strength. Lay conceptions of blameworthiness, with all their nuances, inform the practical application of the law. One result is the ironic constraint upon prosecuting the blameless when dealing with strict liability offences.

Moral evaluations are tied to the exercise of choice. They are made about the activity associated with pollution and the extent to which the actor's deed represented a real choice. What prompts condemnation is not so much how much damage was done, but how much an actor chose to break a rule. The moral judgment is, however, also divorced from the act of polluting and instead reflects on the degree to which a choice is made to comply or not with the rules. In both cases, the stat is symbolically significant, despite the fact that in pollution (as in most strict liability offences) the legal sanctions are not severe, because the ceremony is mutually recognized as one normally reserved to signify condemnation of a bad choice.

IV. Law As Last Resort

This final Part focuses on the role of the formal legal process in pollution control. Chapter 9 analyses the ways in which potential cases for prosecution are pared down in organizational handling, and suggests some of the pressures moulding the number and choice of cases to be prosecuted. The threads of the book are pulled together in Chapter 10. Its main themes are summarized and some of the issues of the opening chapter revisited. The conclusion suggests why the water pollution control authorities (and, indeed, other regulatory agencies) enforce the law in the way they they do, and why, in particular, they make such sparing use of the formal powers of prosecution conferred on them by law.

9. Using The Law

I. The Process

Use of the formal law is the culmination of the enforcement process in pollution control.[1] Prosecution is employed in only a very few cases of those pollutions which theoretically could be brought to court.[2] In two consecutive twelve-month periods roughly coinciding with field work, the northern authority, which prides itself on its reputation for toughness, mounted 21 and 8 prosecutions respectively. The low prosecution rate and the extent of organizational filtering prior to prosecution is evident in the table.

Table: Prosecutions in a pollution control agency

	processed as 'pollution incidents'	formal samples taken	prosecutions mounted *	percentage of 'pollution incidents' prosecuted
Year 1	2212	172 (7.8%)	21 (12.2%)	0.9%
Year 2	1895	123 (6.5%)	8 (6.5%)	0.4%

(source: agency documents)

* Percentage is of formal samples taken

These data are interesting as a description of organizational behaviour, useless as an index of pollution. Field men themselves exert substantial control over the kinds of incident which they define and report, thereby formally admitting a pollution to the system of control for processing as an 'incident' (see ch. 5). The number of 'polluting incidents' should in no sense therefore be taken as an indicator of the 'real' or 'hidden' level of pollution. *Such an indicator is impossible to produce.* As the table shows, few pollutions are serious enough to warrant a formal sample, and few prosecutions follow in those cases in which formal samples are taken—an average of about ten per cent over the two years. Fewer than one per cent of 'pollution incidents' are prosecuted in the northern authority, even though it is engaged in a campaign to clean up some badly polluted rivers in its territory. The southern authority makes still less use of the formal legal process. Over the past five years about seventy formal samples have been taken annually, with about three prosecutions a year (private agency communication).

Use of prosecution is both an outcome of an organizational process and a reflection of moral judgments. Yet it is such an exceptional sanction that, though threatened at field level, officers—and many senior staff—are largely unfamiliar with the process. The legal rules are rarely in the forefront of an officer's mind. In contrast with the police, the pollution control officers' organization provides no incentives for them to mobilize the formal law: in a compliance system organizational outputs are vague and resist quantification, and promotion is not contingent upon a display of successful enforcement as measured by arrest or clear-up rates. Few field men have ever been involved in more than a handful of prosecutions;[3] those that have are the ones who work in urban and industrial districts where a policy of cleaning up rivers has been pursued. Some of the younger men have never had a case prosecuted, for all their vocal enthusiasm for a less conciliatory approach to enforcement. Law is familiar only as the backdrop against which all the field officer's activities are organized: the foundation of his authority and ultimate source of coercion over an unwilling polluter where gentler and more discreet forms of social control have failed.

Prosecution is a visible endorsement of legal rules and is typically engaged in where certain features in a case demand a public display of enforcement and sanctioning (see ch. 10). The infrequency with which the formal law is enforced, however, given the large number of pollutions which theoretically could be prosecuted, gives rise to a high degree of selectivity in its use. Avoidance of the process is to be observed at personal and organizational levels. The following remark may be taken as the epitome of a compliance strategy:

'I think that prosecution is just our strongest weapon, and it might well gain strength, gain usefulness, by being kept as a last resort. ... The objective of the job is not to maximize the income to the exchequer by getting fines. The job is to make the best use we can of the water for the country. ... We get more co-operation if we keep prosecution to the last resort.'

The use of prosecution in a compliance system signals the collapse of a conciliatory approach. For this reason it is common to find pollution control staff, as well as others (such as probation officers) whose enforcement work is directed toward compliance,[4] taking the view that to have to mount a prosecution is a confession of failure. It is at once an admission that the strategy has broken down, a betrayal of the personal failure of their negotiating skills, and a suggestion that a compliance strategy may have been based on faulty premisses. Though it is much to be preferred as a means of enforcing the law, 'the difficulty is that negotiation, without enforcement of the law, can be most protracted,' as an officer with a reputation for a particularly conciliatory style put it. But sometimes that point, the culmination of an approach based on incremental enforcement, is reached when dealing with persistent problems, and the agency has to display its authority: 'We're a law enforcement agency at the end of the day.' In enforcement of traditional crimes, in contrast, prosecution is an item of organizational output to which the police attach great value.

The prerequisite for almost all prosecutions is the legal sample. Once taken, agency staff have a decision to make as to whether it should be followed up by prosecution or set aside while the field man tries again to negotiate compliance. As the table above shows, about 90 per cent of formal samples fall into this latter category, either having been taken as 'frighteners' or because field men felt constrained organizationally to adopt a failsafe procedure and cover themselves, or because headquarters did not back up the man in the field.

A decision to prosecute is the culmination of a series of recommendations originating at field level and handed up through the agency hierarchy. The chain of decision making is rather protracted. The process begins with informal consultation between the field man responsible for the formal sample and his supervising officer. Though these consultations will for organizational purposes be recorded and transformed into a report of the polluting incident and the field man's response to it, the supervising officer relies heavily on his field man's oral account. Recommendations as to whether or not the agency should pursue prosecution are typically embedded in both oral and written reports:

'I've told my [field men] when they write their report to say whether they consider prosecution should go ahead or not, whether they in fact took it only for that purpose. So the decision as to whether to proceed starts with me reading the [field man's] report. I mean, I've already been told about the incident verbally anyway by then.'

In one of the agencies (the procedure is not signficantly different in the other) a report of the formal sample taken, once it has been served and the report written, is routinely transmitted to the head of pollution control. The area supervisor concerned is consulted for advice as to whether or not a prosecution should be mounted, and intermediate officials may also be involved in the discussions. Cases which are at this stage earmarked for prosecution are then referred to the agency's legal department for advice on problems and technicalities which may weaken the prosecution case in court. If there are none and it is agreed that prosecution is appropriate, the case is then passed for final adjudication to an advisory committee. Since the head of the agency during the fieldwork period was also a member of the advisory committee it would appear that he possessed considerable authority in formulating the final decision.

The vocabulary of the reports of pollution to be considered for prosecution going to the agencies' advisory panels is one of negligence and persistence. It is also one of newsworthiness; sometimes the number of complaints received is mentioned; occasionally it will be observed that an incident 'has received extensive coverage in the press'. In fact agency policy about the selection of cases for prosecution is consistent with the view in the field (the only difference being one of numbers suitable for prosecution). Suitable cases are pollutions 'caused by gross neglect or carelessness, or wilfully', where 'a material depreciation of the value of the receiving water has occurred', or those pollutions where such depreciation will occur 'if remedial action is not taken by the polluter, and the polluter is not using his best endeavours to put acceptable remedial action in hand' (unpubl. agency document).

Though the decision about prosecution passes through several hands and is rather remote from the field ('the sharp end'), a field officer does possess some control over the ultimate agency decision. He is in close contact with his area supervisor whenever he takes a formal sample since that act announces to the agency that it has a serious case on its hands. Direct access to his supervisor gives the field man a personal opportunity to explain informally, and, if need be, forcefully, why the stat was taken. Substantial disagreement about the merits of prosecution between field and area men is unlikely. Most area men (and, indeed, most senior staff) are alert to the organizational principle to support the man in the field:

'Field staff see all the pollutions occurring—that's all they do see—and it's very depressing as far as they're concerned to see what are obviously illegal discharges taking place with no action being taken. And so I think, if only to maintain staff morale, one has to make sure that in the serious cases——that some action is taken.'

Besides, having once been field officers themselves, supervisors tend to proceed on the same lines as their junior colleagues in assessing appropriate action. The area man comes to his conclusion

'from a sort of a set of base rules ... like, y'know, "Is it a first offence?" and "Was it an accident?" and all the rest of it. I mean I formulate an idea anyway as to which way it ought to go, whether there is anything to be achieved or anything to be proved, and this is borne out in the [field man's] report, because he usually ends up by saying ... "I only took this to speed them up and as soon as I took it they were OK" and we won't proceed. So, I mean he and I really do the major filtering.'

A sensitivity to a notion of equity of treatment is one such 'base rule' often imposing a real constraint. 'I don't think it's morally right for me', said an area man, 'to prosecute ... a company, say, for ... a BOD of 30 when the consent condition is 20. I know that half our sewage works are probably doing just that. ...' If a stat has been taken as a frightener, or as a covering tactic, the field officer is able to display it to his supervisor as a case not to pursue. Where matters are 'totally beyond human control, for example, a power failure ...' to quote a hypothetical case discussed by one field officer where he would be obliged to take a stat, but a prosecution would be unwarranted, 'I would take the stat and then have it rejected' (an interesting expression revealing the conflict between organizational and moral values).

Where, however, there are differences of opinion between field and area men as to which cases are most appropriate for prosection, there is usually a preference for fuller enforcement at field level, in keeping with the pattern of administrative decision-making in legal bureaucracies where the commitment to a more legalistic approach is stronger in the lower reaches of the organization (Kagan, 1978:161; Robison and Takagi, 1968; Weaver, 1977:100). Recommendations not to prosecute which originate at input and middle-management level—that is with field officer or area supervisor—tend to be accepted at senior level almost always: it is recognized that taking a stat as a 'frightener' is a familiar tactic in persistently troublesome cases. On the other hand, recommendations supporting prosecution, usually the major source of disagreement between field and area men, are more frequently overturned as they proceed up through the organization. Despite a general desire to stand by their field men, some area supervisors play a role in curbing what they see as an occasionally excessive enthusiasm among some of their field staff (particularly some younger officers) for prosecution. In doing so supervisors attempt to impose a measure of consistency upon decisions originating from their area, just as senior staff attempt to exert control over their area men. These organizational relationships make for a degree of equilibrium in agency policy on use of the legal process, restraining any individual extravagance: 'If I decided I'm going [to go] out and belt ten industrialists,' said an area man, 'it would cause a squeak at the top.'

The extended process of agency decision making takes time, allowing polluters to make efforts in the meantime at repairing problems and otherwise displaying signs of willingness to comply. Some polluters are not tempted to take remedial action until a Notice of Intention to Commence Proceedings is delivered to them. Though this may occur many weeks after the pollution (a trial normally taking three to six months to arrange, sometimes more), their efforts may well be enough to persuade the agency to call off a prosecution. For agency staff the time is a chance for further reflection, particularly upon the moral components of the case, which sometimes leads to a decision to drop a prosecution. Even though the pollution grows remote by time, place, and person, compliance is a continuing thing, constantly offering the polluter opportunity to repair harm done. With moral evaluations and designations central to the determination, a practical show of remorse by the polluter usually leads to a recommendation not to prosecute. After all, the agency does not wish to appear unnecessarily authoritarian in public and to prosecute for a problem which has already been remedied may suggest that it is either being unnecessarily vindictive or it is wasting taxpayers' money. The rationale most frequently used—'If the offence is over and done with, what's the point of prosecution?'—reveals the characteristic concern for ends and is not one which would be recognized by enforcement agents in a sanctioning system. Prosecution of the now-compliant deviant becomes unwarranted:

'If I think there's not much point in prosecuting—they're contrite, it wasn't particularly the fault of the management as such, they've taken all steps to prevent it happening in the future, they've been responsible in that they've instantly reported, they didn't try to cover it up, they've taken all steps that were required ...—then you might get to the point of thinking "What's the point of prosecution? Is it going to achieve anything in the future?"'

The instrumental rationale for prosecution is now thrust into the background, exposing its expressive—punitive—counterpart, and prompting anxieties in the agency about the appearance of vindictiveness. It is for this reason that in one authority 'practically all prosecutions ... are for serious "one-off" pollutions' (private agency communication).[5]

II. Organizing Data

Processing cases in an organization offers opportunities for the introduction of bias to the selection of cases for further action. Decision making in legal settings is a social process.[6] This is most explicitly so where decisions are made in consultation or interaction with others, where, for example, decisions are the outcome of deliberations by panel or board. But social processes are equally crucial in those cases where 'decisions' are the product of a referral system in which information is compiled and supplied at one level in an organization to those in whom authority to decide is formally vested.

Even the pollution control officer, the most junior agency official, is in a position to negotiate the designation of cases as 'worth prosecuting' or not. As gatekeeper to the system of legal control he initiates the whole process, and as such possesses some power to manage the outcome of any particular case by exploiting his position as supplier of information and assessments to his seniors. The field man does not effectively have to compete with other accounts in reconstructing the pollution. It is he who defines acts and events, characterizes individuals, and compiles 'information' in the production of an ostensibly objective report.

Any gatekeeper inevitably creates and transforms data. In formulating reports and descriptions of pollution incidents the officer defines the situation for his superiors. Sometimes creation of data is achieved unwittingly with claims to report 'facts' or write 'objective' accounts. Sometimes it is recognized that accounts have to be truncated and there is scope in writing reports for the expression of personal values. And sometimes it is artfully done by those who are well aware of their role in the organization as controllers of descriptions and evaluations and who exploit their position as suppliers of knowledge. They consciously design their reports in an attempt to influence the discretionary judgments of senior officials.[7] Some measure of control over organizational decision making is important for field staff where their continuing need to preserve workable or credible relationships with polluters for the future must be protected by the selective application of prosecution. Similarly, in cases where the field man is handling an obstinate polluter it is important to him to be able to prosecute where he feels the interests of enforcement are best served.

To influence outcome requires a subtle approach from lower-level staff in the way events and individuals are portrayed and assessments made. An apparently dispassionate account to which is appended a statement of personal opinion is regarded as effective in presenting a point of view. The seeming objectivity of such a report commands attention, where blunt, but blatant, language would be discrediting. 'When one writes the report it has to be factual and pretty coldly put out,' said one field man

> 'because it forms the bones of your statement and you can't say, "We ought to do the bugger because he's a right sod," sort of thing, because that would be typed in your statement that's read out in court. You've got to present the facts as you saw them, the impact it was having on the watercourse, the reaction of the offender and your considered conclusions and opinions. But they have to be sensible and dispassionate. You can obviously add a memorandum to the report expressing your feelings more strongly if you feel it appropriate.'

Others, however, prefer to offer their own evaluations subtly concealed as an integral part of the account: 'I, when I write up my report about a stat, I try to put certain inflexions in the report.' Or, as another officer said, less delicately: 'I just bend [my reports] as they go up.' Or, ingenuously: 'When you write up a report a slant's going to creep in.' One field man, recognizing the all-pervasive significance attached in the agency to moral evaluations about the cause of any pollution, suggested how the dependence of senior staff upon accounts and assessments produced at field level could be exploited. Since all but the gravest of accidental pollutions are considered undeserving of prosecution, 'If a pollution is completely accidental, it won't get into the report ... whereas if it's obviously negligent ... we'll make something out of it. In fact we are discriminating at that stage. We have to do it very safely so we don't come in for criticism.'

Another kind of bias is introduced at senior levels in the agency in the course of processing candidates for prosecution. One of the concerns of headquarters staff in potential prosecution cases is the protection of the public image of the authority. The authority must not be seen to lose a prosecution: 'generally ... our line is that we have a cast iron case,' said an agency head, 'and it's only some quirk of justice that would cause us to lose it.' The fear of losing has a major influence on the formulation of prosecution policy; like other regulatory bodies,[8] pollution control agencies, as an area man put it, will 'go to court only on a sure-fire winner'.[9] This is little consolation to field staff who wish to sanction intransigence: 'the authority', runs a complaint often heard from officers, 'needs to be 110 per cent certain of winning a case before it goes to court.' This aspect of prosecution policy is subject to very tight control, as the following snatch of conversation with a senior official who played the major part in his agency's prosecution policy suggests:

SO I never play anything unless I'm going to win.
KH ... How many cases have you lost?
SO I don't think we've lost any, you see.

Criticism of the agencies' prosecution policy from field level is common. Though all field officers are committed in varying degrees to a strategy of compliance, almost all believe their superiors to be too cautious in the use of prosecution, a view stemming from the experience of having had recommendations for prosecution turned down. This caution is customarily explained by both headquarters and field staff as follows: since field men are 'at the sharp end' they see issues surrounding prosecution in a narrow context whose boundaries are defined by the day-to-day demands of managing individual relationships within a geographically modest district. This particularized knowledge fails to provide the field officer with a framework to understand the decisions of his seniors or the magistrates. Senior staff, in contrast, are more influenced by factors exterior to the law's ostensible concern (the present act of pollution) and this wider context creates a sensitivity among them to the broader implications of the use of the agencies' power to prosecute.[10]

Senior staff regard the agency as in an essentially vulnerable position where use of the ultimate sanction is concerned. It is ostensibly committed to a deterrent rationale in its use of the formal law, even though the penalties at its disposal are low.[11] It must therefore only prosecute punishable cases. 'We wouldn't like to be put in the position of taking somebody to court who gets an absolute discharge,' said a senior man. 'This would obviously be a lot worse than not going in the first place.' From field staff's point of view this deterrent rationale argues that losing a prosecution is a public display of agency impotence which will encourage others to take less care about their own pollution control arrangements. As important is a fear expressed at senior levels that a failed prosecution threatens to establish a precedent that the agency is capable of losing: at present a very small proportion of defendants plead not guilty, but any future defendants will learn that a 'not guilty' plea can pay off. Polluters as a result will become more obstructive, by being less co-operative in general and by pleading not guilty in particular.

The cautiousness and reluctance which characterize agency prosecution policy also derive in part from the fact that once the agency's relationship with the polluter passes from the informal, low-visibility setting of negotiation to the formal, public arena of the courtroom, ultimate control over the handling and fate of the case also passes from the agency. Agency staff regard this shift as a transfer of control from fully informed expertise about pollution work, to a lay panel of magistrates, ill-informed and innocent outsiders. Anything less than a watertight case may induce the polluter to plead not guilty, thereby exposing a court ignorant of both the scientific aspects and hazards of pollution and the 'real facts' of the case to misleading arguments in mitigation mounted by the defence. A measure of uncertainty about outcome or about sanction is now introduced which the agency, with its preoccupation with deterrent efficiency and maintenance of its image, cannot tolerate. Concern for image also means the agency cannot risk the appearance of malevolence which may follow an unsuccessful prosecution. In both pollution control authorities, therefore, the case which satisfies the criteria of the prosecution policy in terms of its gravity and blameworthiness will be closely screened by agency lawyers to ensure that there are no impediments of a technical legal kind relating to the evidence which may result in the prosecution being lost.

For field staff these concerns are a further source of interference in the protracted sequence of recommendations about prosecution. Preoccupied with the day-to-day business of control and nearer what they consider to be the 'real problems', they become critical of what they regard as a total unwillingness on the part of their seniors to gamble in cases which they believe deserve or need prosecution. To lose occasionally is not thought to be counter-productive. Complaints are sometimes heard, however, that senior staff are too concerned with protecting the authority's image, maintaining their own personal relationships, making concessions to the local economy or, generally, 'playing politics'. The stock of knowledge available in the area office can supply occasional examples of cases in which powerful individuals had allegedly been able to persuade or pressure the agency to call off a prosecution.

The contrasting perceptions reveal the different concerns of field staff and their superiors in headquarters. The field officer wants to maintain an image in negotiations with individuals that he is a legitimate and authoritative legal actor. Though a prosecution may fail, at least it has been brought, and some punishment is inherent in the criminal process—whatever the outcome. Senior staff, however, are much more sensitive to the image of their agency as a whole, its larger public, and the need to display (to different segments of the public) the fact that the agency is reasonable in the exercise of its power.

III. Images of the Law

Field officers only present themselves as authoritative legal actors when it is tactically appropriate to do so in negotiation. In the routine business of sampling, advising, and bargaining with dischargers, the law is an entity whose existence gives the field officer an ultimate purpose and authority, but one which is distant, dimly-perceived, and little understood. The officer, with few exceptions, has scant knowledge of the precise law he is administering and enforcing. Such knowledge as he possesses is conceived of in terse and simple terms. The practical demands of the job do not require a broad grasp of the legalities and illegalities of pollution (though the *appearance* of knowledge is valuable). Some officers carry copies of the relevant statutes around in their cars, but this is as much to impress polluters that the field officer is a legal enforcement agent and pollution control a matter for the law, as it is a resource for extending the officer's knowledge and clarifying his understanding of the rules he is enforcing. Routine work demands no more than this.

Dischargers are also generally unfamiliar with and somewhat indifferent to the precise law (Brittan, forthcoming). All but the largest companies (which involve their own legal staffs in pollution control matters) are regarded as ignorant of the legal rules and penalties.[12] Dischargers will normally accept what is ostensibly a statement about law from a field officer as authoritative, the field man needing fear no challenge to the accuracy of his pronouncements.

The rarity of prosecution means that most field staff have little direct contact with the legal process or the legal profession. Yet their impressions of the formal process—and even some senior staff's—are stark, being drawn, like other aspects of the job, from personal experience. Their images have been garnered from exposure to particular prosecutions in which they or their colleagues have been involved, and from other sources through which they acquire a more general knowledge of what law and lawyers are like. Stereotypes abound. Law is regarded as an arcane speciality, unnecessarily technical and fussy in its dictates, unpleasant in its formal processes. Thus to collect evidence which will stand up in court involves enormous care and sometimes imposes impossible demands on the field officer. Legal procedure is a hindrance, preventing certain kinds of knowledge being used during a prosecution and impeding the just sanctioning of a 'good case'. The pragmatic perspective of the officer treats certain evidence as relevant to an understanding of a polluter's behaviour and problems. It is frustrating, for example, to be balked by the rule excluding evidence of the polluter's past performance, for it is often this, as much as a particular incident, which is a reason for bringing a polluter to court. The 'true facts' of a pollution are thus frequently suppressed by the demands of legal procedure.

Though they urge a higher prosecution rate in principle, the actual prospect of tangling with lawyers is not relished by field men. Lawyers are seen as clever individuals, who, when employed by the other side, will 'shoot you down in flames' or 'tear you to pieces' if your account of the pollution is remotely open to challenge. Courts are not, therefore, pleasant places.

Problems with lawyers, officers often believe, are exacerbated by magistrates who preside in pollution prosecutions. Like other enforcement agents (Skolnick, 1966:196-7), field staff regard the courts 'as remote from, and unsympathetic to, the real problems of enforcing [the] law' (Cranston, 1979:117). Where pollution control is concerned, magistrates are regarded as ignorant laymen, possessing neither the knowledge or experience of field staff, ignorant of the causes and treatment of pollution, and lacking the technical and scientific awareness to make informed decisions. Their inability to distinguish between pollutions of greater or lesser seriousness leads them, it is thought, to punish many polluters with derisory fines.

The low fines meted out for regulatory offences have been explained by some as perhaps due to the 'general identification of magistrates with high-status offenders' (Paulus, 1974:110).[13] It seems, however, that in sentencing pollution offences magistrates tend to behave as they do when sentencing other kinds of offender; that is, they fix penalties in relationship to the maximum permitted by statute. And in those cases where the sentence does appear low by the standards of the tariff, provoking criticisms that magistrates do not really 'understand' pollution, the divergence of view can be understood as a product of two discrepant images of reality. One is the magistrate's knowledge of the case, which is limited to the evidence produced in court in support of a prosecution. This consists of brief, technical accounts of one or more polluting incidents shorn of their history and context. It is presented (as field men see it) by a contrite defendant now guilty of an isolated aberration which he has made every effort to remedy. The other image—what the field man 'knows' to be the case (cf. Wheeler *et al.*, 1968b:48-9)—is very different, and purged from the prosecution account. In both major incidents and persistent failures to comply, the incident and the polluter are set in the context of a long past history. With persistent failures to comply and many one-off cases, it will be a history of misconduct, unco-operativeness, and non-compliance which provides significant meaning and relevance to the lengthy but ultimately unsuccessful set of negotiations which prefaced the polluter's prosecution. Magistrates see a cross section of reality, and sanction it accordingly, where enforcement agents have a longitudinal view of career.

Field men's impressions coalesce to create a broad view of the formal legal process as an alarming and unsatisfactory ceremony, in which the available knowledge about a pollution relevant to sanction—the 'real' knowledge—is submerged beneath the weight of procedural controls on what may be put forward in open court. Ironically, given their enthusiasm for greater use of prosecution (compared with their seniors), it is not, then, a matter to be embarked upon lightly. The officer's willingness to press for prosecution is affected by not only the contingencies of pollution control work, but also the constraints imposed by the nature of the legal process: involvement in a prosecution takes time and causes trouble. The procedural niceties require an extra commitment on the part of the field officer, demanding a sacrifice of routine—of organized, familiar, comfortable work. In marginal cases where the officer cannot make up his mind whether or not to take a stat, or to recommend prosecution, the view of the court as an awful place may even be decisive.[14] It is ironical that very few polluters indeed offer not guilty pleas, thereby opening the issue of the pollution to adjudication by the court. It is only in these circumstances that a field officer might expect to have to enter the witness box.

IV. Postscript

When serious or persistent pollutions are dealt with as potential prosecution cases, organizational values intrude themselves upon judgments made in the field. One such value is the *posture of invincibility* which the agency seeks to maintain when putting its behaviour on display in the courtroom, for it is this which makes the agencies credible as legal authorities. The implication of this concern for senior staff's decisions about cases to take to court is that the question they address is not whether there is a prima-facie case, but whether there is a certainty of conviction, illustrating the proposition that the 'rational component of formal organization avoids the fortuitous, the random and the contingent ...' (Blumberg, 1967:61). Though a strict liability law is intended to facilitate prosecution by dispensing with the need for evidence about the state of mind of the accused, the need for a competent vindication of the agency's authority means that it will never risk going to court with a prosecution that might be lost. From the point of view of legal policy the result is that there is very little opportunity for administrative practice to be scrutinized by legal values.

In a compliance strategy, it is prosecution which is, for practical purposes, the final stage in the enforcement process, not conviction and sanction. In contrast with the police which organizationally have a stake in attaching the label of criminality to offenders, the pollution control agencies are interested in compliance. Their preoccupation is with clean water, not convictions. The great reluctance to employ the formal process coupled with the great likelihood of conviction mean that the crucial decision is not what happens in the courtroom, but rather which case is selected for prosecution. The taint of the legal process comes from an association with its machinery and appearance in court, not from the punishment it metes out.

10. Law As Last Resort

I. Prosecution *in Extremis*

A central concern of this book has been the role of law in regulation. My intention has been to portray the formal process of prosecution as a kind of *éminence grise*, a shadowy entity lurking off-stage, often invoked, however discreetly, yet rarely revealed. Why is it so little used and by what principles are particular cases selected for prosecution? In general, what explains enforcement behaviour in this arena of legal control?

Though the water authorities only employ prosecution *in extremis*, this is not to suggest that law is not central to their concerns. Law provides the mandate and the context for agency activities. And field staff themselves are legal actors, implicitly defining every day the reach of the law. Practical criminal law—the enforcement of the norms embodied in that branch of the law—is, to paraphrase Matza (1964:176), founded not so much on the substantive acts it deems unlawful, but rather on principles that define its proper realm and procedure. In pollution control the conditions under which formal intervention is deemed to be morally and organizationally permissible are very narrowly construed indeed. The irony is that the adoption of strict liability does not expedite formal enforcement of the law, for taking advantage of strict liability is regarded as being 'unreasonable': '[The agencies' predecessor authorities] were expected to act reasonably by prosecuting only for flagrant and careless breaches of consent conditions.' (Agency document.)

Discussions about the role of formal enforcement in regulation are, in general, of two kinds. Behind both lurks an implied complaint about a rate of prosecution that is too low. One argument is couched in economic terms. Resources are inadequate, it runs, with the apparent implication that agencies would prosecute more vigorously if only they could afford the personnel or time (e.g. Cranston, 1979; Conklin, 1977). Thus it has been claimed in the United States that the 'greatest handicap to the successful enforcement of agency regulations' is limited agency budgets and inadequate staff (Clinard *et al.*, 1979:35).

Financial impediments, it is held, drive agencies into a conciliatory posture. The enforcement of standards against individual polluters, 'may be quite inefficient in terms of time, money, and results, and bargaining with violators may often be necessary' (Davies, 1970:176). This suggests an economically rational view of prosecution, one which ignores the moral component at its heart and the social context in which enforcement is conducted. The allocation of resources does, of course, impose its own ultimate constraints on what an enforcement agency can do (Long, 1979), but there is no evidence to suggest that the water authorities' present prosecution rate is one inhibited by lack of resources. Indeed, the director of one agency suggested that a doubling of its annual prosecution rate could be absorbed without any noticeable impact on the work of its legal department or senior staff, while the head of his legal department claimed he could 'mount three or four times the amount of prosecutions ... without any difficulty'.[1]

Another group of explanations is framed in terms of the relative power of the regulator and the regulated. Capture theory is the best known example of this species. In another version, the low rate of formal enforcement is testimony to the power of corporations (Nader, 1970:viii). But there is little evidence from the present research to support this argument, whether or not it is true of the American position. Many polluters comply, or try to comply, generally as a matter of principle; besides, the water authorities are in a rather strong position when deciding whether to prosecute, and possess a virtual monopoly of the information on which a prosecution will be based. Despite their apparently modest legal penalties the agencies enjoy a variety of practical powers which are seemingly rather effective in the personal encounters where enforcement work is done. Large companies, indeed, were frequently portrayed by field men as nervous of agency authority. Little reliable evidence came to light to suggest that such companies attempted to use excessive muscle or influence upon the agencies beyond that which would be considered appropriate in establishing a negotiating position.[2] Here again, the field man's typical portrait of industry emphasized a defensive posture.

Another version of the explanation centred on power emphasizes social status.[3] This suggests that the reluctance of regulatory agencies to prosecute more vigorously can be understood in terms of deference to social status (Mileski, 1971). Agencies tend to process high-status offenders by administrative rather than legal means, the argument runs, particularly where complainants are of low status. This sort of explanation may be relevant where agencies have identifiable complainants as third parties in the enforcement relationship (as in housing code violations) but there is little evidence of its validity in pollution control work where complainants are seldom involved directly, and where those open to prosecution possess no kind of homogeneous social status.

In this chapter I want to put forward a different view about the strategies employed in enforcing regulation, emphasizing the role of formal legal processes. In essence, the argument is that regulatory enforcement is a symbolic matter, reflecting intimately the conjunction of privately-held (but shared) values with organizational interests in enforcing a secular code of conduct about which there is a high degree of social and political ambivalence. I stress again that the discussion is about regulatory enforcement strategies in general, even though for convenience it takes the place of prosecution for its focus. In presenting the argument I shall also recall the main themes of the book.

Any treatment of the problem must address two questions: first, what prompts the level of prosecutions mounted? Secondly, what are the properties which mark out certain cases as worth prosecuting? In considering the former question I shall rest the argument on an analysis of prosecution as a public act and on a recapitulation of the features promoting a compliance strategy. The more elaborate discussion on the interconnected question about the selection of particular cases deemed deserving of prosecution necessarily involves a measure of overlap but for clarity is dealt with separately in discussions of efficiency and justice.

Why is there such sparing use of prosecution? The pollution control agencies, like other regulatory bodies, operate in a public environment. They are public bureaucracies, publicly funded, exposed to a measure of scrutiny, and vulnerable to public criticism. Their audience and their image matter. No organization exists in a vacuum, as Selznick (1966:10) reminds us. 'Large or small, it must pay some heed to the consequences of its own activities (and even existence) for other groups and forces in the community.' Thus, as an area man put it, 'the preservation of public confidence in the authority is a relevant factor in the consideration of legal action'. As public bodies the water authorities are sensitive to the conflicting demands of the two competing and potentially critical constituencies representing the 'environmentalist' and 'business' perspectives which I outlined in Chapter 1. In their use of the formal legal process the agencies express the nature of their relationship to their environment, while the divergent views represented by the constituencies themselves suggest the extent of the ambivalence surrounding the conduct to be controlled. Selecting cases for prosecution is in large part a consequence of the organizational need to manage appearances before these two ill-defined, but none the less real constituencies with an interest in agency work. The necessity is to satisfy the demands of the environmentalists for action, while achieving the quiescence of the business constituency.

Enforcement work in pollution control, like police work, is bound up with the management of appearances. Prosecution is a symbolic act, 'the dramatization of the moral notions of the community', as Thurman Arnold (1935: 153) put it. It isolates a target group, 'shows it to be vulnerable, and implies that the problem it represents is controlled by such symbolic action' (Manning, 1977:248). Prosecution is visible evidence of the commitment of an agency to its legal mandate, for, in the words of a senior man, 'the authority's not seen to be doing its job unless it does prosecute'. However lowly the setting, however feeble the sanction, prosecution for pollution is a ceremonial statement of what is desirable in the public interest (Gusfield, 1967). The agencies feel exposed to public attention in the exercise of their authority—to endorsement of their policy or to the risk of public criticism of bullying or extravagance.[4] Prosecution transforms the enforcement of regulation from a private ordering of relations between the polluter and the agency in which the nature of control has been shaped by bureaucratic and moral constraints, into an open contest based on the principles of formal legal control, in which the agency publicly seeks to establish and sanction the polluter's rule-breaking. For these reasons the agencies have a strong incentive to prosecute only those who are likely to offer guilty pleas, thus merely reserving the question of sanction for independent adjudication.

Prosecution also brings regulatory deviance to public view, modest though the publicity of the local newspaper or the trade journals might be. Publicity makes possible the vindication of the agency as a credible enforcement authority. Public enforcement visibly displays regulatory rule-breaking as the law's business, dramatizing the success and effectiveness of the agency, and enhancing such deterrence as resides in the criminal process. *The need to dramatize a measure—but a carefully controlled measure—of activity is the more acute given the lack of substantial consensus about the agencies' mandate.* For this reason the large majority of the pollutions which agencies handle—the routine cases—possess too few symbolically significant features to make them worthy of public enforcement (Manning, 1977:249). The issue here, ultimately, is one of organizational self-preservation, for organizations are very human institutions which behave in very human ways. 'Prestige and survival are not normally accepted as legitimate ends of administrative behavior,' Selznick writes (1966:65). '... But prestige and survival ... are real factors in decision. ...' The management of appearances towards potentially critical constituencies imprints agency prosecution policy with a concern for self-preservation.

Yet the agencies lack workable indices of impact and success which may be conjured up to display their effectiveness. Success in regulation in fact is often less visible than failure (Diver, 1980:274). A policy of fuller enforcement expressed in a swelling prosecution rate cannot be employed as an indicator of efficiency, for in an environment of ambivalence this risks being treated as evidence of agency harassment. It is, rather, in the careful and sparing selection of cases for prosecution, that the agency is best able to protect its own interests by showing that 'something is being done': 'MPs, local councils, anglers, members of the public, pressure groups, from time to time have to see that the authority is doing something. ... And if it's not doing what it should be doing then it's quite deservedly open to criticism.'

As I argued in Chapter 1, the environment of regulatory control is one riddled with ambiguity and ambivalence. The very notion of 'regulatory justice' is itself the embodiment of ambiguity, while ambivalence about the nature of regulatory control is central to the adoption of strategies of compliance. Legal rules are not generally enforced in court because the regulators and the regulated have a mutual interest in not enforcing them in this way. Agencies paradoxically believe their efforts to attain the wider goals of their legislative mandate to be *facilitated by the extensive (formal) non-enforcement of the specific offences*. The degree to which an agency turns to the formal procedures of the law depends on its perception of the relative strengths of its significant constituencies. Since the perceived social and political environment is a shifting thing, regulatory agencies may be expected to shift accordingly in the extent of their commitment to a strategy of compliance on the one hand, and a sanctioning strategy on the other.

In pollution control work discretion is moulded by an imperative to conciliate, for officials seek to put problems right or to prevent recurrence of damaging or hazardous events. Compliance implies continuity. The open-ended enforcement relationship creates a context in which an officer judges individuals, acts, events, or incidents. His discretion, informed by such continuity, and exercised over a small and stable segment of the population, is expressed in longitudinal perspective, on the basis of what is 'known' over a period of time and what might need to be done in the future.

Pollution control work is adaptive, given to serial enforcement by negotiation, and to incrementalism, with the gradual application of pressure and, if necessary, ever more ominous threats. Flexible control is made possible by time, bargaining, and privacy. A range of tactical options is available. Earlier decisions may be revised, and the timetable of compliance stretched or tightened. For the deviant's part, even scant display of remedial efforts may be enough morally to foreclose *for the time being* the possibility of sanctioning on the grounds of unco-operativeness or non-compliance. To display willingness to work towards conformity with legal standards frees the deviant from imputations of blame. But such a compliance strategy stems from the symbiotic character of the enforcement relationship and is only possible where there is little recourse to law.

These features serve in fact only to emphasize the similarities in the behaviour of the police and regulatory officials. Where the police deal with states of affairs (cf. Bittner, 1967b) and with a familiar population (Gardiner, 1969; Whyte, 1943; Wilson, 1968) their enforcement in less serious cases also takes on a conciliatory character:

> The officer emphasizes non-legal solutions to problems rather than legal ones as he progressively becomes more involved in interactions with the people with whom he has to deal. A working norm for the uniformed officer in situations such as these is perhaps best expressed as limited enforcement, arrest is used only in cases of flagrant violations. [Petersen, 1968: 206-7][5]

Similarly, the pollution control officer confronted with a flagrant pollution incident is likely to think in terms of a penal rather than a conciliatory response. A strategy of compliance is only possible where rule-breaking is not weighty. In all but the most massive pollutions, it is the component of moral disreputability—of wilful or negligent rule-breaking or persistent disregard for the enforcement agent's authority—which can make deviance serious.

In pollution control and other regulatory enforcement work the behaviour subject to control is also significant in another sense.

Pollution treatment processes involve chemical, biological, and technological applications whose functioning and maintenance are sometimes only imperfectly understood. Unless negligence or malice are readily apparent, deviance and control in this context tend to be regarded by enforcement agents—at least at the outset of a relationship—as essentially 'technical' or 'scientific' in character, lending themselves readily to a judgment that they may lie beyond the immediate practical control of the discharger, and sometimes beyond his economic capacity. Often harms are not readily determined and victims are diffuse. This again prompts a compliance strategy. But where suggestions of culpability or lack of co-operation enter the picture, however, the problem, once more, is much less likely to be designated as 'technical' or 'inevitable', for here matters are most certainly regarded as preventable. Blameworthy conduct is punishable conduct.

In the more familiar areas of behaviour embraced by the traditional criminal law, compliance usually means refraining from an act. But in pollution control compliance often requires a positive accomplishment, sometimes with major economic implications. Time and money have to be spent in one form or another, in planning, buying, building, and maintaining compliance. Money is not always immediately available; planning and consultation can be time-consuming; equipment is often long in the delivery; facilities are not quickly designed and built; erratic treatment processes take time to settle down. To insist on instant compliance is an affront to moral sensibilities about what is reasonable. Compliance in practice usually means time, for compliance is not often, and cannot normally be, instant:

> 'you've got to give people time to do it. It's not "You mustn't do it today, and if you do it again tomorrow I'll smack your legs," you know. It's not like speeding. ... I can instance a discharge in [one town] that is not consented that's been discharging for two hundred years. ... The factory's been there for two hundred years and making this discharge for two hundred years. You can't suddenly leap in—it's not like a motoring offence. ...'

The result of all of this is that pollution control staff must display patience and tolerance, rather than legal authority, for their goal is not to punish but to secure change. After all, the only alternatives to graduated compliance are often more appealing in theory than in practice, for they involve immediate stoppage of a discharge (which could shut down a factory) or disposal of the effluent elsewhere (which with anything but a very small volume discharge could impose crippling costs). Since both alternatives have far-reaching implications for employment and the economic position of a company, common sense at field level dictates that such measures are beyond the polluter's control.

If the consequence of all this is a working presumption against use of the formal processes of law, in what kinds of case will prosecution actually be employed?

II. Efficiency

Strategies of control are shaped by the continuing need for what enforcement agents regard as effective pollution work. A notion of efficiency is of central concern at both field-officer and agency level, acting as a constraint against too ready a commitment to a sanctioning strategy. Legal bureaucracies, like others, are preoccupied with issues such as the maintenance of internal and external relationships and the management of work. These concerns have major implications for the kind of justice dispensed. Highly selective use of the formal processes of law is implicit in regulatory control founded upon compliance since to use prosecution in any but the most serious cases is regarded as counter-productive. Negotiation is the effective way of achieving results.

The thoroughly pragmatic approach to pollution control work shared by field men is expressed in the primacy accorded to 'getting the job done' through the maintenance of relationships with their client population. The high discretion each officer enjoys is operationally efficient; he is the only person who 'really knows' the dischargers, their problems, and their negotiating styles. The field man's conception of effective work calls for the active co-operation of those whose behaviour he is regulating, and the cultivation of good relationships eases the task now and in the future. Though a legalistic view of regulatory enforcement might contemplate the use of prosecution as an efficient tool for an enforcement agent (presumably serving the ends of deterrence), I have stressed that field staff in practice regard it in all but a few cases as obstructive (cf. Macaulay, 1963). It is inimical to the attainment of compliance, the overriding goal of regulation, and incompatible with effective discovery and detection: 'The trouble is, once you've prosecuted somebody it leaves a sour taste in the mouth, and it's very difficult to ... be on good friendly terms with them thereafter. ... [T]hey don't forget it, and you've lost that influence—that friendly influence—that you had before.' Or, as agency policy has it:

Experience shows that a general strict application of the letter of the law is not conducive to effective pollution control. Practical realities [are] to be taken properly into account. ... [I]t is generally more important that pollution should be prevented and recurrence avoided, than it is that polluting acts be penalized ... [Unpubl. agency document.]

The result, in the words of an agency head, is that '... the authority determines a low level of enforcement, and it's really that it has at the back of its mind that this is the *best* level of enforcement, because you can upset people. You can get so much done by not upsetting people.' (His emphasis.) To prosecute a case in which the polluter does not acknowledge the agency's interpretation of his behaviour will be seen as undeserved and resented. Regulatory law in these circumstances does not intimidate so much as alienate. If prosecution is to be employed, therefore, it must be with a high degree of selectivity. Sparing and judicious use of the power is further underpinned by recognition that going to law is disruptive to the essential work of the organization. 'Of course prosecutions are a lot of work,' said a senior man. 'Normally we don't do any more than we have to. And it's largely unproductive work so far as pollution prevention is concerned.' In enforcement of regulation, then, the handling of routine cases is informed by organizational norms of efficiency. It is in the exceptional case where the norms of justice deserve conspicuous display.

In the field, intransigence may normally be avoided by practice of the officer's art of 'common sense', eschewing too precipitous a shift to an adversarial stance which threatens his control over the private management of enforcement. Negotiation confers control over the rules of the game: the timing and choice of tactics are the officer's, and the grounds on which compliance is defined and sought are his also. The adaptive nature of negotiating helps preserve the semblance of the control which is a major preoccupation of enforcement agents committed to a compliance strategy, and which, when competently managed, almost always pre-empts the disruption of prosecution. There is virtually no opportunity—since the stage is hardly ever reached—for the deviant to escape the legal sanction, to 'get off'. It is precisely when a deviant appears to be 'getting away with it' that recourse to the law is had, the ceremony of the law only entering the picture when the enforcement agent has exhausted the tactical possibilities and has effectively lost personal control: 'We don't take people to court just like that. It's a history of problems. We've tried everything with them: negotiation, discussion, etc. When we take them to court, it's like saying all the other methods have failed.' In these cases field staff have nothing more to lose. Where too ready a resort to the law would once have been counter-productive, the deviant now presents himself *in a way recognizable to himself and others* as deserving of legal sanction for his obduracy. It is not the continued deviance which prompts the response, so much as the persistence which signifies wilfulness. The non-compliance is transformed by it into wrong-doing of the same order as the deliberate pollution incident, where field staff feel they have everything to lose by not prosecuting. The persistent rule-breaker deserves sanctioning for his intent, constructive or actual. Here compliance strategy yields to a sanctioning strategy.

A relationship based on bargaining lends itself to a certain stability, assisting efficiency by enhancing predictability and the organization of behaviour. While efficiency at field officer level is expressed in terms of a desire to negotiate and protect relationships, at agency level it is transformed into an imperative to portray the authority as all-powerful. Where, in routine cases at field level, efficiency is a constraint against use of the formal process, the most serious incidents of pollution are a matter for general agency concern. Here efficiency postively demands prosecution. Such incidents possess features which make prosecution appear a rational and reasonable response, even—very occasionally—in the apparently complete absence of culpability. Prosecution here not only furthers general deterrence, it also serves to display publicly the agency carrying through its legal mandate as a credible enforcement agency, an important recognition of the interests of its environmentalist constitutency in salient cases. Furthermore, such a show of strength may help convince those polluters whose inclination is to comply that they will not be commercially disadvantaged by their deviant competitors (Kagan and Scholz, 1979).

But agencies, like individual field men, are also reluctant to yield control over the handling and disposal of a case. The high degree of selectivity in the cases which go to court reflects in part anxieties about the surrender of control to the lay judgment of magistrates who will dispose of the case by principles which are by no means similar to the principles employed in everyday administrative behaviour. If the agencies cannot play on their pitch, by their rules, with their referee, they can at least choose their opponents and decide whether or not to go through with the fixture. The caution is striking, given that the agencies have a strict liability law at their disposal. To lose a case, however, is held to be extremely inefficient from a deterrence standpoint (see Rabin, 1972). Such a prospect is too severe a blow to contemplate. In surrendering control over the adjudication and disposal of a pollution prosecuted, the agency is at the same time yielding to procedures which create a salient case. However much the odds are stacked in their favour prosecution remains a gamble for agency administrators. Losing the gamble risks proclaiming to the agency's clientele that ultimately it is not omnipotent. The mystique of the law and the legal process cultivated at field level risks exposure, for there the threat of prosecution may no longer be the trump card.

Achieving what agencies perceive to be optimum efficiency in use of the formal law demands a balance be struck. Some degree of formal enforcement is necessary for symbolic purposes. After all, a weapon which is never used may provoke doubts about the will of those who possess it to use it:

> Doubtless, enforcing some rules may have a self-defeating property. However, most rule enforcement persists as a practical option because it is useful for enforcement agencies. [It protects] law enforcement organizations from the appearance of laxity. By setting more severe public symbolic standards, they may be able to attain modest levels of control or deterrence ... [Manning, 1977:248.]

From the pollution control agency's point of view, to sum up this part of the argument, there are two types of case where efficiency decrees prosecution to be the most appropriate response. First, where persistent, noticeable failure to comply is concerned, belief in individual and general deterrence demands public sanctioning. The formal legal process is employed here to enable the agency and its field staff to preserve their credibility for the conduct of future negotiations. Secondly, where a pollution incident which causes substantial and noticeable damage, hazards water supplies, or involves the agency in heavy expenditure takes place, the response will again be to prosecute. Here the agency is compelled publicly to declare the fulfilment of its legal mandate. Given a sufficiently grave pollution, this will be the case even if there is no evidence of blameworthy behaviour on the polluter's part (as in the chromate pollution described in Chapter 8), for the scale of the pollution itself will help shield the agency from the appearance of vindictiveness. On the contrary, prosecution will be demanded: 'It should be borne in mind that [in] special cases of occurrence, or grave risk, of massive pollution damage, or of great public outcry, the taking of a prosecution may be necessary, irrespective of the other practical circumstances of the case.' (Unpubl. agency document.) Not to be seen to be taking action against such a striking breach will invite criticism that the agency is failing in its duty, while the sheer gravity of the pollution will serve to mute any complaints from the business constituency: 'OK if there's ... a lot of damage, no-one seems to mind very much.' Such cases possess particular potential for dramatizing the agency's prowess as an enforcement authority.

But it must be a grave pollution for such action to be taken against the blameless; agencies are extremely reluctant to announce publicly the moral status of one who may not be blameworthy, for the use of the apparatus of criminal trial is intimately bound up with the suggestion of moral disreputability. The gravity of the harm is the key factor in this kind of case, outweighing any complaints that morally blameless behaviour is being stigmatized as criminal by the enforcement of a strict liability statute. Such cases demand a symbolic display of agency authority. *It must act in 'big' cases.* The environmentalist constituency will expect action, while its opponents can hardly object in the face of such conspicuous harm. Thus it was entirely appropriate for the water authority which prosecuted the company in the chromate pollution case to take the very unusual step of having it tried in the Crown Court, where greater public attention to the proceedings could be expected and where legal sanctions were higher. Senior staff expressed considerable satisfaction with the outcome of the trial, and were also delighted with the fact that BBC television cameras had been filming stretches of the river on the day of the prosecution in preparation for a news story on the case.

III. Justice

Throughout the book I have argued that moral judgments play a central part in the use of enforcement discretion. Where agents deal with persistent problems and compliance is delayed, moral inferences may be drawn of the polluter's responsiveness to the enforcement process, leading to sanctioning of the unco-operative on the grounds that they had a choice in the matter. A prosecution here represents a failure of compliance strategy, and is needed for purposes of credibility and deterrence. Where specific acts or incidents of pollution are concerned, however, moral evaluations of the act are of crucial importance in establishing whether it involves blameworthy behaviour by the polluter worthy of sanction. The moral content of the act makes any prosecution deserved, for this rule-breaking resembles more familiar forms of traditional crime. Here prosecution acquires an expressive character. In these circumstances blame is crucial in the public enforcement of the law, for to display publicly an indifference to the question of intent would, in Holmes' words, 'shock the moral sense of any civilized community' (1881:50; also Matza, 1964:71).

There is little explicit moral content relating to intent in the pollution control legislation, in keeping with most offences crudely categorized as 'strict liability'.[6] Yet although in the setting of standards equity is a guiding principle, and in their enforcement moral judgments are all-pervasive, there is in practice, paradoxically, a marked ambivalence about use of the formal machinery of criminal law to sanction pollution. The values society seeks to protect in regulating economic activity occupy a morally problematic position, as the existence of the two competing 'environmentalist' and 'business' constituencies suggests. The police, in contrast, are often portrayed as enforcing norms about which for the most part a high degree of social consensus exists. Traditional crime is a symbolic assault upon a society's fundamental moral integrity demanding repressive enforcement. Pollution—and doubtless other forms of regulatory misconduct—enjoy no such status.[7] Indeed, the behaviour of regulatory agencies should be understood as a response to the lack of consensus about the values society wishes to advance (Stone, 1975:97).

In seeking to promote social interests, regulation burdens productive behaviour. It imposes costs on industrialists and agriculturalists which must ultimately be borne by the public, while constraining activities which are not widely regarded as morally beyond the pale. In pollution control work, field men deal with behaviour which is possibly 'illegal' but hardly ever 'criminal'. The ambivalence is also reflected in the pollution legislation (and, indeed, in other forms of regulatory legislation) which defines conduct as criminal in strict liability terms but provides for rather modest amounts of punishment (in contrast with those sanctions available—and sometimes meted out—for breach of traditional criminal law). In regulation we see a compromise: the criminal law is given a wide potential reach but little sanctioning power. Strict liability offers ready enforceability, but the penalties impose little cost. Yet in practice, the perception of an illegal act as a 'problem' rather than a 'crime' leads to a substantial proportion of cases being defined by regulatory agents as outside the proper province of criminal law (cf. Clinard, 1952:298; Lane, 1954:94-5). The reluctance of officials to link much regulatory misconduct with blameworthiness is a persistent feature:

Violators of the laws the agency administered were looked upon as honest businessmen who had inadvertently engaged in practices that conflicted with some of the complex legal regulations. The agent was expected to understand their predicament and to help them correct their mistakes. To be sure, violators were ordered to cease their illegal practices, but only a very small proportion of them—typically wilful, repeating offenders—were brought into court to be penalized for having broken the law. [Blau, 1963:190.]

It has been argued that such reluctance is partly due to the recency of the legislation carving regulatory deviance out of behaviour which used to be perfectly legitimate. The implication of this, so far as breaches of the Factories Act are concerned, is that 'Thieves, the penal system's closest equivalent to the alleged white-collar criminal, violate rules of conduct which, whatever their origin, are by now a traditional part of our culture; the factory-occupier seldom contravenes such rules even though he may break the criminal law.' (Carson, 1970b:397.) In effect offences against regulation have not been culturally absorbed and do not invite the same condemnation as breaches of the traditional code—except where hazardous behaviour is the result of negligent or deliberate misconduct.

Use of the formal law is heavy with meaning for the pollution control agency. Criminal trial stigmatizes, setting up an internal constraint against extensive use of prosecution, since the agencies and their field staffs will not countenance prosecution in any but the most blatantly culpable cases; a more legalistic policy of enforcement, a senior officer said, would lead to the courts being 'full of morally fairly innocent people'. In effect those moral features in any pollution case which make prosecution undeserved, serve in traditional crime merely to mitigate punishment. But there is also an external constraint, as a significant segment of the public is regarded as intolerant of stigmatizing economically productive activity as criminal, except where clearly blameworthy behaviour is involved:

'We're dealing with legislation in matters which the general public doesn't *really* consider to be part of the criminal law. ... They don't consider themselves criminals because they've discharged an effluent containing one part per million more BOD than they ought. ... And also ... because these minor transgressions of ... applying for consent don't in fact, in many cases, actually *do* any damage. If someone hasn't applied for consent, what he hasn't done is comply with an administrative requirement of the Act. But he hasn't in fact *done* anything. ... So in fact to *prosecute* him for not having applied for the right piece of paper in this country doesn't go down too well. And so——I think this is the real thing, that people just don't consider that they are criminal acts. ...' [His emphasis.]

To enforce the law against the blameless is to diminish the moral authority of that law.

But where blameworthy behaviour is concerned, the regulators' moral mandate can be made clear. Thus law is used where enforcers can rely on what they perceive to be a consensus of values. The legal process here is a morality play, a symbolic vindication, censuring blameworthy conduct and preserving shared social bonds. It is this concern for culpability which forecloses the possibility of significant criticism from the anti-regulation constituency. *What is really being sanctioned is not pollution, but deliberate or negligent law-breaking and its symbolic assault on the legitimacy of the regulatory authority.* This is where the agencies see a consensus. In securing their salient enforcement activity in a recognizable framework, regulatory agencies are anchoring their behaviour in pervasive, deeply-held norms.

Commonsense conceptions of criminality shared by the general public and enforcement agents alike hold blameworthiness to be the essential ingredient in a decision to invoke the law, hence the emphasis given to selecting those cases which have violated commonsense principles of justice. While being highly selective about prosecution, field men are still enforcing a code of behaviour which somehow corresponds with their perceptions of the moral sentiments of the community. Most pollution cases yield few instances of unambiguously blameworthy behaviour. At field level, pollutions are not often defined as being the result of calculated or purposive behaviour or indeed the consequence of a deliberate unwillingness to spend money to comply. While agency staff regard their standards as generally attainable without extravagant cost, causing pollution is viewed more often as the inevitable consequence of physical impediment, limited economic means, or the result of carelessness or inefficient management, or of accident. And where there is a suggestion of culpability it must be clear-cut, with noticeable consequences. 'If it's a relatively minor offence,' said an older officer, 'it goes against the grain that people should be prosecuted or fined.' Furthermore, the technological problems involved and the complex industrial setting in which many pollutions take place often make it very difficult for an agent satisfactorily to establish cause and allocate blame. Hence it is a relatively small proportion of culprits for whom the formal process and its stigmatizing consequences accord with the agency staff's moral view of just deserts. In formulating and enacting prosecution policy senior staff are extremely sensitive about avoiding the appearance of vindictiveness. As one senior man put it: 'Public authorities—the big risk on publicity is that they will be castigated in the press as the big heartless bureaucracy victimizing the private citizen.' The risk is particularly acute in compliance systems where a problem has been corrected in the time necessary to mount a prosecution. Negotiating an outcome to a problem need not be concerned with such questions.

Two kinds of culpable misconduct invite the regulatory agency's ultimate sanction. In serious one-off pollutions where officials can be satisfied that the incident was the result of culpable misconduct, prosecution will be regarded as deserved. 'Desert' is a concept very carefully construed in practical terms by the water authorities. While desert may be relevant in private transactions at field level in shaping routine responses, it must be handled with restraint if there is a chance of prosecution, because public sanction almost always means public condemnation. However, there is a nagging uncertainty in potential prosecution cases as to the target for public condemnation. While an unequivocally blameworthy regulatory offender becomes indistinguishable from the traditional criminal when convicted and punished, there is a risk that in the absence of clear desert the condemnation will be directed towards his accusers.

In visiting their personal conceptions of justice and desert onto a formal law which is indifferent (so far as cause is concerned) to *mens rea*, administrative officials inevitably narrow the field of incidents deserving of prosecution. Though I have emphasized the importance of efficiency in structuring enforcement behaviour aimed at securing compliance, the demands a field man makes of a persistently non-compliant polluter and his response to intransigence—the other kind of culpable misconduct—are ultimately also geared to a moral appreciation of the deviant's co-operativeness. There is a clear preference to take only 'obvious cases' of moral disreputability. If a polluter does not see himself at fault or the public cannot infer fault from the damage done and what it knows about the discharger's behaviour, the prosecution of that discharger may again be the occasion for public criticism. *Such constraints demand prosecution of 'the bad' cases:* the malicious, the negligent, and the conspicuously obdurate who have resisted the legitimate efforts of the agency to enforce its legal mandate. Here, again, the agency can be confident of expressing shared values, for the most fervent opponent of regulation can hardly complain about the vindictiveness of an authority which prosecutes those who flout the law. Industrialists, it seems, share the same values as agency staff as to what warrants punishment (Brittan, forthcoming).

IV. Postscript

The formal processes of the law will be employed only where a regulatory agency can be sure they rest upon the secure foundation of a perceived moral consensus. In the vast majority of cases of regulatory deviance a confusion of interests and values exists, manifested in doubts about whether agencies are protecting the public good when sanctioning behaviour which is a consequence of economic activity beneficial to the public. *In an environment of ambivalence what is mutually recognizable assumes immense importance.* Conduct placed within a moral framework possesses just such recognizability. The formal machinery of the law is appropriate therefore only in those 'big' and 'bad' cases where the gravity or moral offensiveness of a breach renders the confusion of interest minimal. The agencies are demonstrably doing something while offending few. This is deeply-entrenched behaviour in pollution control work and perhaps a major reason for the persistence and consistency of the patterns of enforcement behaviour to be observed in a wide variety of areas of regulation. To regard a compliance strategy of enforcement at field level, or the formulation of prosecution policy at headquarters as symptomatic of the capture of the regulators misses the point. The practice of regulatory enforcement expresses an identity of moral values which transcends the regulator-regulated relationship.

While rules are enforced by human beings, the rule and its breach will always be set in their social contexts, leading to judgments about desert or equity. Regulation may be contemplated by the law as the dispassionate sanctioning of misconduct by the even-handed application of a criminal law unconcerned for the niceties of *mens rea,* but regulation in practice, mediated as it is by a bureaucracy in which people have to exercise their discretion in making judgments about their fellows, is founded upon notions of justice. Pollution control is done in a moral, not a technological world.

Notes

Chapter 1: Introduction

1. Frankel, 1974; Goldstein and Ford, 1971; Gunningham, 1974; Irwin, 1970.
2. Glenn, 1973; Hines, 1966a; 1966b; Holden, 1966; *Texas Law Review*, 1970.
3. Cranston, 1979; Paulus, 1974; Silbey, 1978; n.d.; Silbey and Bittner, n.d.; Smith and Pearson, 1969; Steele, 1975.
4. Carson, 1970a; 1970b; Geis and Clay, 1980; Kelman, 1981; Law Commission, 1969; Stearns, 1979.
5. Carlton *et al.*, 1965; Mileski, 1971; Nivola, 1978; 1979; Ross and Thomas, 1981; (but cf. Gribetz and Grad, 1966).
6. Blumrosen, 1965; Mayhew, 1968.
7. Beaumont, 1979; Clinard, 1952; Kagan, 1978; Thompson, 1950.
8. For example, Stjernquist, 1973 (forestry); Schuck, 1972 (meat inspection); Truman, 1940 (agriculture); and see generally Dickens, 1970; Lidstone *et al.*, 1980. While compliance has also been the dominant style in enforcing regulation in the US (Anderson, 1966:70), it has recently been argued that a more punitive approach is now observable (Kagan, 1980 and private communication). This trend may well turn out to be short-lived with the advent of the Reagan administration (*Washington Post*, 1981).
9. There is a large political science literature in which this theme appears: see, e.g. Fesler, 1942: Herring, 1936; Huntington, 1952; Leiserson, 1942. Perceptive recent analyses are by Sabatier, 1975, and Wilson, 1975; 1980a. Critiques of the notion by political scientists have recently emerged: see Katzman, 1980, and the papers in Wilson, 1980b.
10. It is tempting to follow recent work by Reiss and Biderman (1980) and Reiss (1980) which I have found particularly valuable and employ their vocabulary of 'compliance systems' and 'penalty systems'. Since they are unlikely to subscribe to all of the ideas subsequently elaborated in this discussion, however, I think it advisable to adopt slightly different terms (albeit those employed by Mileski, 1971).
11. I use the word not to describe the behaviour of enforcement agents (which can be a good deal less than conciliatory when circumstances demand) but to suggest that the style is one which refrains from full and formal enforcement.
12. I am grateful to Albert J. Reiss jun. for clarifying some of the ideas in the following paragraphs.
13. The word is not used normatively. I use the terms 'deviance', 'rule-breaking', 'misconduct', and 'violation' synonymously throughout.
14. Hence the evident willingness of the authorities to prosecute for television- or vehicle-licence evasion (Lidstone *et al.*, 1980).
15. A penetrating analysis of serial enforcement is presented by Rock (1968; 1973b), some of whose terms I have adopted.
16. e.g. Banton, 1964; Bittner, 1967a; Chatterton, 1979; Cumming *et al.*, 1965; LaFave,

1962; McCabe and Sutcliffe, 1978; Parnas, 1967; Punch and Naylor, 1973; Wilson, 1968.

17. I follow agency usage throughout in occasionally employing the plural.
18. This filtering is analogous to the processes in civil legal actions (see Ross, 1970). The pattern is also observable in traditional criminal cases, but the scale differs; prosecution is a much commoner response in cases entering this system of control.
19. For convenience I speak in general of 'the polluter' which suggests an individual and unfortunately obscures the fact that field staff for the most part deal with organizations; see ch. 7.
20. Failure to treat ambivalence as a problem in its own right, rather than a mere definitional difficulty, has bedevilled much of the literature of white-collar crime (Aubert, 1952). See, e.g. Burgess, 1950; Hartung, 1950; Sutherland, 1945; Tappan, 1947; and generally Geis and Meier, 1977.
21. See Diver, 1980. A good illustration of the point is provided by the Wisconsin water pollution control authority which avoided enforcement when the local major industry of cheese making was in a slump. Manufacturers were going out of business and the agency wished to avoid the political ramifications of the suggestion that pollution regulation was contributing to economic hardship. The decision was made 'to let nature take its course before [stepping] in with pollution requirements for the survivors' (Murphy, 1961:124).
22. It is important to emphasize that I am caricaturing these constituencies from the agencies' perspective.
23. Persuasive historical analyses have argued that in some areas the business constituency has promoted regulation. These views are most commonly attributed to Kolko, 1965; 1967; but see also Carson's interesting paper (1979) and the debate it provoked (Bartrip and Fenn, 1980a; Carson, 1980a).
24. The words are from a private communication from Peter Manning to whom I am grateful for prompting some of the ideas in this section.
25. The term is not wholly satisfactory but it at least avoids the possibility of misunderstanding, which Matza's term 'consensual crime' does not. Becker (1970), incidentally, uses the term 'conventional crimes' to refer to *mala prohibita* because they are seldom subject to public criminal sanction.
26. See below pp. 18-19. The picture is rather different in the US (Morris, 1972) where penalties for regulatory offences have been stepped up in recent years (Clinard *et al.*, 1979:31). Perhaps the best known example of the fearsome sanctions which are possible in America is the notorious pollution of the James River by the pesticide Kepone, discharged by the Allied Chemical Company, which met with a fine of $13,240,000. Furthermore, many forms of American regulatory prohibitions have civil as well as criminal penalties, which presumably serve as intermediate sanctions. For example, two-thirds of the US income tax authorities' resources are allocated to enforcement activity, of which only ten per cent are devoted to criminal, rather than civil, enforcement (Long, 1979). See generally the helpful essay by Thomas (1980: esp. 110-11).
27. This seems to be typical of other areas of regulatory control: for instance, offences under the Factories Acts were punished in 1967 by fines averaging £33 (Law Commission, 1969:23; cf. Bartrip, 1979; Bartrip and Fenn, 1980b). Imprisonment, however, was sometimes used to penalize repeating adulterators of food in the early years of this century by using the charge of 'obtaining money by false pretences' (Paulus, 1974); this suggests that the conduct concerned was regarded

as possessing properties which aligned it with traditional crime.

28. Though this is less true of the position in the USA. A broad view of trends in that country is presented by Thomas (1980).
29. See below. The House of Lords addressed these issues in the case of *Alphacell* v. *Woodward* (1972). LORD WILBERFORCE and VISCOUNT DILHORNE employed a justification for strict liability based on expediency of proof, the former referring to the unnecessary 'complication' of *mens rea* (ibid. 479), the latter arguing that if Parliament had intended any mental element to be a prerequisite in causing pollution, 'a burden of proof would rest on the prosecution that could seldom be discharged' (ibid. 483). LORD SALMON preferred a public welfare argument, holding that it was 'of the utmost public importance that our rivers should not be polluted' (ibid. 490). He went on to place pollution with other examples of regulatory offences, describing them, in the words of WRIGHT J., in *Sherras* v. *De Rutzen* (1895:922), as acts which 'are not criminal in any real sense, but ... which in the public interest are prohibited under a penalty'. Furthermore, LORD SALMON continued, if the appeal were to succeed, 'a great deal of pollution would go unpunished and undeterred to the relief of many riparian factory owners' (ibid. 491).
30. Predictably, where the police do deal with the locals they tend to be conciliatory in the interests of not making enemies (Preiss and Ehrlich, 1966:21-2).
31. Compliance, however, is a more complex matter than it may seem: see ch. 6.
32. To suggest that a measure of compliance exists runs counter to much of the prevailing rhetoric in America, where it is customary to condemn regulation as having failed to secure the compliance of businesses.
33. e.g. Banton, 1964; Bayley and Mendelsohn, 1968; Bittner, 1967a; 1967b; Black, 1980; Black and Reiss, 1970; Bordua, 1967; Cain, 1973; Chevigny, 1969; H. Goldstein, 1963; LaFave, 1965; Manning, 1977; 1980; Muir, 1977; Niederhoffer, 1967; Petersen, 1968; Piliavin and Briar, 1964; Reiss, 1971; Rubinstein, 1973; Skolnick, 1966; Westley, 1970; Wilson, 1968.
34. Reiss (1966) made a plea for studies sensitive to the role of organizations in a classic paper. Things may now be changing at least in America, thanks perhaps to the impact of consumerism and public disenchantment with big business. Recent work which may help redress the balance includes Edelhertz and Overcast, 1980; Ermann and Lundman, 1978; Reiss, 1980; Reiss and Biderman, 1980; Schrager and Short, 1978; Shapiro, 1980.
35. Trade effluent pollution control is described by Richardson *et al.* (forthcoming), who also discuss the structure and reorganization of the water industry at greater length. On the latter, see further McLouglin, 1976; Sewell and Barr, 1977.
36. Some discharge to underground strata.
37. The authorities may prosecute under other Acts, but in the context of this research these are of little importance.
38. The research was conducted at a time of legal change. The redesigned apparatus of control was in place (Water Act, 1973), but many sections of the Control of Pollution Act 1974 were awaiting implementation. The offences and sanction levels in force during fieldwork were those laid down in the Rivers (Prevention of Pollution) Acts of 1951 and 1961.
39. See n.27 above.
40. The fine was up to £50 for every day the offence continued, or £500 (whichever was the greater).

41. Or a fine of £10 per day or £100 (whichever was the greater), or both.
42. An agency term and one used advisedly, since all the officials involved in the research were men.
43. With the exception of part of the southern authority, where staff also have other responsibilities, and are not under the direct control of headquarters.
44. Though the field officer in the northern agency reports regularly to his area office, he works from his home. His area supervisor, however, will normally expect to spend more time at the agency's central headquarters with his other area colleagues and the senior administrative staff than in the area office.

Chapter 2: Setting Standards

1. See further Ackerman *et al.*, 1974; Davies, 1970; Richardson *et al.*, forthcoming.
2. It follows from this that the law may be irrelevant as a means of control or sanctioning unless the standard actually addresses 'pollution' specifically. What is visibly 'polluting' to a layman may legally be clean water. In one case, for example, a major water supply and fisheries river was contaminated by the escape of a large amount of chemical dye which ironically was normally used to trace the source of pollutions. As a result the water in the river for several miles turned conspicuously red. The chemical could not be treated, but simply passed through sewage works without effect. The dye did not affect the suspended solids, biochemical oxygen demand (see n. 10 below), or any of the other parameters laid down in the company's consent. The water authority in this case was theoretically powerless: there was no standard at the works where the pollution originated which addressed the colour of its effluent, therefore no pollution.
3. Consistent demands made of dischargers in the standards set, apart from expressing a deeply-held value of justice, also assist in efficient enforcement, Kagan (1978:81) argues, since rules which are uniformly and consistently enforced are more likely to meet with voluntary compliance.
4. An unpublished agency document reported that 'only about 50 per cent of samples of existing discharges conform with currently applicable quality conditions. ...' Subsequent chapters discuss the reasons for such official tolerance.
5. There is a parallel here in the reluctance of regulatory agencies to pursue prosecutions where defendants are likely to contest the charge strenuously: see, e.g. Cranston, 1979:125.
6. Solids and ammoniacal nitrogen exert oxygen demand; metals may be toxic, cyanide in sufficient concentration lethal. Thermal pollution affects fish.
7. Applications for consent must address the nature and composition of the discharge, its maximum temperature and volume, and its highest rate of flow. A consent cannot be unreasonably withheld; it may be reviewed.
8. Field staff in the southern agency now play only a minor part in standard-setting.
9. These hark back to the Royal Commission on Sewage Disposal 1912.
10. BOD, an index of the biodegradable matter in water, is a measure of the amount of oxygen a given body of water will consume in five days at a temperature of 20°C.
11. i.e., 10 BOD, 10 suspended solids.
12. The suggestion that 'expertise' is linked with 'trust' perhaps indicates the extent of the belief that compliance is a substantially technical matter. In the USA, however, there seems to be a marked distrust of big business and it is unlikely that

enforcement agents would take the same view.

13. Names of polluters and officials are fictitious throughout the book.
14. The 'stat' is the formal sample necessary for purposes of prosecution. A number of different terms are used in both agencies: 'formal sample', 'legal sample', 'tripartite sample', 'sealed sample', 'official sample', 'prosecution sample', and 'statutory sample' (or, more commonly, 'stat'). They all refer to the same thing and are typically used interchangeably in conversation. The officers who have the most extensive experience of legal samples usually talk of 'taking a stat'; the process is described in ch. 8.
15. The charges made to firms to discharge their liquid wastes into the foul sewer for treatment at the local sewage works: see Richardson *et al.*, forthcoming.
16. See also ch. 6 for a brief discussion of efforts by the regulated to co-opt enforcement agents by gifts or bribes.
17. Previously, an individual could only prosecute with the leave of the Attorney-General.
18. Control of Pollution Act 1974, s. 41.

Chapter 3: Field Staff

1. In the western region of the southern authority, however, where there are more assistants and senior field staff and correspondingly fewer posts in the field officer grade, the routine tasks tend to be shared to a larger extent by all three ranks.
2. It must be emphasized that differences of view are not necessarily differences of practice. I was unable to collect any systematic evidence about the extent to which younger staff carried their sympathy for a sanctioning strategy into the field. My impression is that in practice their behaviour does not differ substantially from their older colleagues.
3. This complaint, often directed at their older colleagues, echoes, of course, those writers who have emphasized a 'capture' theory of enforcement.
4. 'Common sense' is also a virtue in police work (Banton, 1964; Chatterton, 1979; Manning 1977).
5. Similar generational differences are to be observed in other areas of regulatory control, e.g. Cranston, 1979: 16; Crenson, 1971: 14-16; Stjernquist, 1973: 149. See also Ross, 1970: 128.
6. Needless to say, 'bureaucracy' is used throughout in a sociological, not a pejorative sense (Blau, 1963:250 ff.; Weber, 1946:196 ff.; 1947:329 ff.).
7. 'A calling is a moral vocation carrying with it extraordinary obligations and prerequisites,' Matza writes. 'A job is just that—a job. It is mundane. Its obligations and rights are ordinary. Among the extraordinary obligations of a calling is a tenacious dedication to one's flock, a noble eschewing of considerations of career, an indifference to the matters of hours worked, conditions of work, and salary, and a deep and publicized belief in the job of one's vocation.' (Matza, 1964:145.) (Matza has presumably borrowed the conception of 'calling' from Weber, 1958.) The pollution control officer does not reveal such a high degree of commitment as that envisaged by Matza. But young and old alike, nevertheless, display a general conception that the work is much more than a mundane job.
8. There were, for example, no complaints from anyone at all about pay.
9. But cf. Carson (1970b:392) and Mawby (1979:247). Another feature of police

work—danger—is rarely present in pollution control work, though there are hints of it in some settings. I was occasionally given instances—some apocryphal, no doubt—of threatening incidents which field men have had to face.

10. The time and emphasis given to different tasks will vary according to the nature of the officer's district and the time of year. Officers in rural areas, for example, will spend much of their time in late spring reminding farmers of the dangers of silage pollution. One of the agencies conducted a time study of the activities of their field staff which showed that pollution control work was comprised on average of the following:

    field time (hours): 12.7 routine surveillance (including sampling)
    8.9 travelling
    4.0 pollution incidents
    2.8 'other activities'
    2.1 planning matters
    0.6 consultancy investigations
    total 31.1

    office time (hours):
    5.7 'other activities' (including telephoning)
    3.3 report writing
    1.6 planning matters
    1.3 letter writing
    total 11.9

    The proportions of samples taken were as follows:
    5.0 effluent samples
    4.6 river samples
    2.8 'others', including groundwater and biological samples.

11. A field man is willing to act as lay consultant, where necessary, in the interests of having a polluter take action. There is an unwillingness to make any kind of commitment in giving advice, however, ostensibly for fear of making the water authority legally liable if it proves unsuccessful. Officers prefer to make 'tentative suggestions' in recognition of the fact that the firmer the commitment to the advice given, the greater the feeling of having underwritten the success of the remedy proposed, and the more 'responsible' the officer will feel if it proves ineffective. In a setting in which negotiation is accorded such prominence, the officer has a professional stake in avoiding this position, for failure discredits and the impression of competence essential to the job will be undermined. Thus while field staff project an image of themselves as knowledgeable about the natural science of pollution control and well-versed in the various treatment techniques, they do not actively play the role of 'expert' in the sense of making precise suggestions as to what kind of remedial measures should be undertaken. Instead, their prescriptions emphasize ends, rather than means. Where there is a real risk of accountability they will tell polluters where they can contact professional consultants.

12. Policemen and other street-level bureaucrats are the same: Kaufman, 1960:134; Manning, 1980:220 ff.; Petersen, 1968:219-20; Skolnick and Woodworth, 1967: 123.

13. What the notes describe will not be representative of the work of all field men; the officer in a less populated area would expect to find fewer pollution problems and his activity would usually be less hectic. What is reproduced is a personal record

and will fail to communicate to readers the other images and memories which they stir. The unrecorded account, as ever, is a pot-pourri of often particularly vivid detail; for example, the amount of time spent driving, even in a relatively compact urban area; the heat in the car and the intermittent crackling of the radio-telephone with its scrambled language and strange one-way conversations; the unexpected symmetry and beauty of Victorian brick-work in one of the city's culverted watercourses, and its hidden world of pipes, doors, drains, and flaps; the din on a factory shopfloor; and the field officer's indignation at the way in which the local paper had, he thought, misrepresented the seriousness of the factory's earlier pollution.

14. Sampling and other monitoring techniques are described in ch. 5.
15. See n. 14, ch. 2.
16. A bund wall is built around a tank to retain its contents should it leak.
17. A classification of water quality according to standards promulgated by the Department of the Environment.
18. The area office is used as a clearing house for the dispatch of samples to the area laboratory, as well as a temporary store (refrigeration is required).
19. The field man must always be prepared to cover for sick or absent colleagues. At weekends staff must stand by in turn to attend to any emergencies which occur.
20. In part of the southern authority there is ostensibly a greater concern for training staff. A deliberate effort has been made to familiarize field men with most districts by rotating the allocation of officers (with two exceptions) to each district, so that each man spends a few months in one district before moving to another. The area is predominantly urban and geographically compact, thus posing few travel or domestic problems. A spell in the laboratory carrying out routine analysis of samples is also included in the rota. The physical arrangement of the area office (open-plan in an attempt to come to terms with cramped quarters) gives field staff more opportunity to learn from one another. Lack of privacy virtually compels them to make their business known to colleagues. Communication between staff in such a setting is inevitable, whether they like it or not: 'We're very open here. Everyone discusses everything with everybody else sort of thing, which probably gets boring for some people at times, but we tend to have a broad idea of what's happening then. I can't honestly see a situation here where anyone would keep their little areas to themselves and no-one would know ... their area. ... Everyone deals with everything in a way.'
21. Stjernquist quotes a Swedish forest ranger (1973:149): 'When I was young ... I took perhaps a harder line and said "This is the law and you must obey it". Then when I grew older and became more experienced I never took this approach.'
22. See n.14, ch.2. It is interesting that these remarks were made by an officer in the northern authority where an effort has been made to crystallize some of the 'normal working standards' in the form of procedural notes. However precisely framed such guidelines to discretion may be, they are often inexact as a practical guide to action in a particular case. One simple example may illustrate the point. One of the procedural notes seeks to instruct staff as to the circumstances in which a formal sample should be taken. One such circumstance is a 'fishkill'. In practice, however, there is a large area of personal discretion as to what constitutes a fishkill. The number and kind of fish and their location were all important features. One dead trout could result in a stat being taken where a hundred dead minnows would not.

23. The 'Rebecca Myth', of which this is an example, is discussed by Gouldner (1954:79-83).

Chapter 4: The Job

1. With the one exception of part of the southern authority noted earlier.
2. Few did—though this may be an instance of the presence of the observer affecting the behaviour of those under study.
3. Schuck (1972) has shown how the acceptance of gifts—'cumshaw'—is an integral part of meat inspection: to refuse a gift makes the meat packer suspicious and less co-operative. The other side of the coin, however, in a setting where the continuing relationship is a central feature of enforcement behaviour, is a potential for future embarrassment should a field man succumb to present temptation (Ross, 1970:64-5).
4. The capacity of organizations to control the activities of their members is addressed by Kaufman, 1960; in connection with policing see Bordua and Reiss, 1966; Chatterton, 1979; Rubinstein, 1973; Wilson, 1968.
5. Since the area supervisor relies on his field officer for knowledge of any particular problem, he tends to have a sense of loyalty to him in dealings with senior staff where there may be differences of opinion—as is often the case where recommendations for prosecution are involved (ch. 9). Area officers seek to maintain the morale of their field staff in recognition of their importance as gatekeepers, for fear that their commitment to the job might be threatened if senior staff too frequently failed to follow their recommendations.
6. The supervisors in the southern agency are based in the area offices from which their field officers work, an arrangement which permits direct contact between supervisor and field staff. In the northern agency area officers are based at headquarters, but most spend two days a week or so at the area office.
7. Towards the end of the research fieldwork, sampling in the northern authority was transformed from a matter very much within the field officer's discretion as to where and when he took samples, to a highly structured programme, in which the number and timing of samples were determined (much to the irritation of most officers) by the demands of the laboratory and the computer.
8. Known in American policing as 'CYA' ('cover your ass'): Johnson, 1972:239 ff.; Van Maanem, 1974.
9. See ch. 5. Though even here, many field men will draw a formal sample to show they mean business, but then pour it away later if the polluter has taken it seriously. Staff can do this in the knowledge that their superiors will not find out: see chs. 7 and 8.
10. Similarly, the detective has an important output measure in the 'clearance rate', whose significant components are the rates of arrest and prosecution (Skolnick, 1966:167-9), and the FBI agent finds conviction rates are important for organizational evaluation (Wilson, 1978:96).

Chapter 5: Creating Cases

1. The concepts of proactive and reactive organization of enforcement have been developed in a line of publications by Reiss and Black. See Black, 1970; 1971; 1973;

1980; Black and Reiss, 1967; Reiss, 1971; 1974; Reiss and Biderman, 1980; Reiss and Bordua, 1967. Katzman's study of the Federal Trade Commission (1980) makes use of the concepts. See also the discussion of 'targeting' by Manning, 1980.

2. Heavy rains in urban and industrial areas wash solids and oil into watercourses, producing a slug of particularly polluted water which limits the amount of natural life any watercourse can support.
3. Similarly, Weaver (1977:81 ff.) notes the reluctance of American anti-trust lawyers to prosecute struggling businessmen.
4. This practice is sometimes endorsed by area men. One said he would 'bend the rules on farms, for example, because farmers are one of the most intractable problems you can get. You can't treat farm effluent satisfactorily. ... So far as I'm concerned the farmers can continue until someone squeals.'
5. Watercourses also have different purifying capacities: fast flowing streams with rapids or falls which allow aeration will purify more quickly than sluggish lowland rivers.
6. One consequence of the different degrees of tolerance shown by pollution control staffs may be that clean rivers become cleaner, while dirty rivers remain dirty.
7. This calls to mind Sacks' observation of the link between deviance and visibility (Sacks, 1972). In many respects noticeability is reminiscent of the policeman's conception of 'outside' (Rubinstein, 1973:341 ff.).
8. Most farm animals are of considerable value.
9. Trout are also a valuable commodity.
10. See ch. 2. The decrease in tolerance, however, may be more apparent than real, being accompanied in many cases by relaxed standards.
11. Summoning up accustomed practice in this way is observable in a variety of legal contexts (see, e.g. LaFave, 1965; Ross, 1970).
12. The Royal Commission on Sewage Disposal 1912 is the source of the frequently found 20 BOD/30 suspended solids standards.
13. BOD—biochemical oxygen demand—is the commonest test for the presence of organic pollution: see n. 10, ch. 2.
14. i.e., 40 parts per million suspended solids.
15. Some police forces have tolerance limits articulated in policy statements stipulating when a violation of the formal traffic law may be ignored: see Petersen, 1968:148-9.
16. pH (a measure of acidity or alkilinity) is a parameter often found in the consents of industrial dischargers.
17. Though where pollution is associated with a lack of territorial location, as in 'cowboy' dumping (see n. 19 below), detection is especially difficult. On this, see Reiss, 1980.
18. This is possible since the agencies had fewer enforcement resources before reorganization and simply could not get round to many of the less obtrusive dischargers. In a recent year in one urban district '35 previously undeclared discharges of a significant volume' were discovered (agency document).
19. There is evidence that some people engage in pollution as a deviant occupation. For example, a trade ('cowboy dumping') has grown up in at least one of the water authority areas in which highly polluting liquids are collected—for a price—and then dumped, not into a toxic waste tip (for which there is a charge) but into a sewer in some inconspicuous location. Consequently, field men are always careful to study the activity going on around any stationary liquid tanker.

20. Field staff's belief that small companies are less concerned about water pollution is supported in a quantitative study carried out in America (Roos and Roos, 1972).
21. Some police officers 'are instructed to develop an attitude of suspicion' as part of their normal training (Petersen, 1968:92-3).
22. Analysis of a routine sample at the time of fieldwork cost about £20.
23. Many American regulatory agencies also publish such statistics of non-compliance—also to the dismay of their field agents, betraying their orientation to results (Robert A. Kagan, private communication).
24. Staff in the northern authority have confirmed this for themselves by finding marked disparities in the results from the same samples of polluted water analysed in different authority laboratories.
25. See n. 1 above.
26. Albeit providing the conditions for self-fulfilling prophecy, with increased surveillance leading to increased likelihood of discovery.
27. The agencies sometimes prepare explanatory leaflets to be circulated by field staff about certain kinds of pollution, such as silage or oil. Their distribution offers opportunities for discreet inspection.
28. Besides, many effluents (especially those from sewage treatment works) may vary in quality from hour to hour depending on load or changes in water use. The agencies themselves prefer representative sample results, another reason for varying the sampling routine.
29. This case also illustrates the preference for proactive monitoring by inspection where possible, rather than by sampling, despite the fact that in the officer's opinion the effluent was well above the suspended solids limit consented.
30. This strategy is well known to the police. Many patrolmen 'break their shift into separate blocks of time with two hours spent in an area known to contain a good possibility for a drunk arrest, two hours in a location likely to yield some traffic violators, and so on. ... An officer who concentrates upon checking for stolen cars will quite often find five or six stolen automobiles a month. As a consequence, he will not encounter many other violations ... ' (Petersen, 1968:98).
31. Thus American police will use traffic stops to question and search suspicious persons (LaFave, 1965:187; Tiffany *et al.*, 1967).
32. This is one way in which the agencies learn of the illegal dumping mentioned in n. 19 above.
33. One of the advantages in being on fishless rivers, one urban officer observed drily, was that he was never phoned up by angry fishermen.
34. Similarly the US Antitrust Enforcement agency has a rule that all Congressional mail must be answered within 48 hours (Weaver, 1977:62-3).
35. Though see p. 168.

Chapter 6: Compliance Strategy

1. About half the prosecutions brought in the southern authority are for persistent failures to comply (agency communication).
2. Similarly Ross and Thomas (1981:12) report that housing inspectors regard violations as 'chickenshit' or 'important'.
3. The concept of career is treated in Rock, 1973b and Roth, 1963.
4. Stopping (rather than cleaning) the discharge is sometimes the ultimate (but not the immediate) aim with persistent failures to comply: some polluters are

encouraged to recycle waste water or dispose of it to foul sewer.

5. So far as the agency is concerned, the nature of the behaviour to be regulated and the realities of the enforcement strategies adopted mean, in effect, that it is difficult to display 'compliance' or other measures of success or impact in any meaningful fashion (cf. Wilson, 1968:57). River surveys will indicate broad changes in water quality over the years which agency staff may attribute to more extensive and efficient enforcement. But it is extremely difficult to sustain this claim with any precision owing to problems in disentangling enforcement practices from other shifts which have had an impact on water quality, such as changing patterns in land use or changes in public expenditure on sewage treatment; see ch. 4.
6. Similarly, businesses are regarded in the consumer field as essentially law-abiding (Cranston,1979: 29). Principled compliance may be commoner in Europe than the USA, as Kelman's data seem to suggest (1981). See generally Anderson, 1966. Brittan's study of discharges (forthcoming) supports the perception of a majority given to principled compliance.
7. Field staff assume that larger businesses have a reputation to protect and are thus 'better' or 'more responsive', a notion suggested by Lane (1953).
8. Brittan (forthcoming) suggests that officers' assumptions about large companies and nationalized industries are correct.
9. Compare the unquestioning compliance which police can expect in regulating traffic behaviour (Bittner, 1967a:702)
10. Kagan and Scholz (1979:15) observe that legalistic enforcement 'seems to have been a primary factor in stimulating political organization by regulated firms and attempts to attack the agency at the legislative level', while a more legalistic approach by the California Occupational Safety and Health authorities has led to an increased number of appeals (ibid. 14). See also Barrett, 1979; Kagan, 1980; Kelman, 1981.
11. See further ch. 1. These penalties have since been increased by the Control of Pollution Act 1974 and the Criminal Law Act 1977—but the sanctions are still generally regarded as modest by field men despite the provision for terms of imprisonment. Most officers, however, only had a very hazy notion of the availability of this sanction.
12. In the USA, the Environmental Protection Agency has consciously used adverse publicity as an enforcement tool: Gellhorn, 1973:1401 ff. See also Rourke, 1957.
13. See further Brittan, forthcoming. Cases have been known to occur in which shop-floor workers have manipulated the system of internal organizational control to encourage an unwilling management to comply. They have acted in a deliberately unco-operative manner, ultimately forcing the field officer to threaten the company with prosecution. This is normally enough to ensure that management make the resources available to improve their pollution control arrangements.
14. The proportion of dischargers presumed to be utterly unco-operative varies according to the nature of the field officer's patch, with those working in urban industrialized areas expecting more deviance and greater reluctance to comply than those whose areas are predominantly rural. This is despite the view of the farmer as 'troublesome'. The apparent contradiction may possibly reflect the fact that farm effluents, in contrast with many industrial discharges, are all relatively familiar and easily traced. One supervising officer from a largely urban area

suggested that about 70 per cent of dischargers were 'co-operative, willing to comply'; about 20 per cent were 'slow to comply'; the remaining 10 per cent or so he classified as 'having to be forced to comply'.

15. This raises the question of the techniques adopted by field staff for categorizing dischargers into those 'genuinely' reporting a pollution, and those taking unfair advantage of pollution control staff, an issue addressed in ch. 8.
16. A former Chairman of the Illinois Pollution Control Board has observed that 'if no one complied until prosecuted, enforcement costs would surely strangle the program' (Currie, 1975:390).
17. Bargaining is one of the central characteristics of legal processes: see Hagevik, 1968; Holden, 1966; Jowell, 1977a; 1977b; Ross, 1970; Silbey, 1978; Stjernquist, 1973: 149; and generally Gouldner, 1960; Strauss, 1978.
18. See n. 16, ch. 3.

Chapter 7: Negotiating Tactics

1. The analysis here owes much to Rock's graceful study of debt enforcement (1968; 1973b).
2. Similarly, closure of factories appears to be hardly ever contemplated by the Factory Inspectorate (Law Commission, 1969: 55).
3. See the figures in the Table, ch. 9, s.i: cf. Carson, 1970b; Rock, 1973b.
4. Interpretative work is central to the behaviour of policemen on the street and their arrest decision making. The motorist who is abusive is more likely to get a ticket (Gardiner, 1969:148), the youth who appears to make claims to toughness is more likely to be booked (Piliavin and Briar, 1964); see also Emerson, 1969:192 ff.; LaFave, 1965; Werthman and Piliavin, 1967; Westley, 1970; and generally Goffman, 1967: esp. 47-95.
5. Save his warrant to gain admittance to land, which is the size and appearance of a credit card and not often used.
6. It is done by talk, the officer's equivalent of the policeman's 'bullshitting' (Muir, 1977:228).
7. Cf. Sternquist (1973:135) who quotes a forest ranger as saying, 'The great majority [of forest owners] considered themselves to be so skilled in managing their forests that they understood it better than a whipper-snapper like me.'
8. It resembles the 'preaching' of probation officers who picture 'avoiding trouble ... as a decision completely within the delinquent's grasp' (Emerson, 1969:224).
9. This view of much organizational deviance as accidental or inevitable contrasts with the tendency among writers on white-collar crime to conceive of it as largely the product of conscious decisions to break the law.
10. Threatening to withhold benefits, however, is a sensitive matter. Some officers claimed not to employ the tactic on the grounds that it is improper, as a supervisor put it, 'to use one set of legislation to implement another'.
11. Discussing the use of court proceedings in debt collection, Paul Rock has noted that 'A creditor who refuses to employ these modes of enforcement may still try to exploit them. [Creditors] will threaten the debtor with a sanction they have no intention of using' (Rock, 1973b:66). Ross and Thomas (1981) have observed similar bluffing in the enforcement of American housing codes.
12. Lawyers acting for dischargers play a very small part in pollution control work

(presumably in contrast with the position in America). In more than two years' field work I only once attended negotiations in which a company's lawyer was present.

13. This Act is discussed by Brown, 1976, and Dickens, 1974.
14. At the time the fine was £100.

Chapter 8: Practical Criminal Law

1. Taking a stat is analogous to filing suit in civil law: Ross, 1970:215 ff. It is possible for prosecutions to be based on evidence of pollution taken only from samples upstream and downstream of the discharge, which will indicate its impact on the watercourse, but this procedure is, apparently, very rarely used.
2. See the table in ch. 9, s. i.
3. See n. 1, ch. 6. The use of the stat to signal a failure of negotiations parallels the process in civil law dispute processing where the formal machinery is also used as a last resort: see Ross, 1970.
4. 'Nine out of ten' is much more accurate: see the table in ch. 9, s. i. The latter figure is a statistical exaggeration designed to emphasize the fact that the majority of statutory samples taken by many of the officers—particularly (it is believed) the younger ones—are not proceeded with by their superiors. Some of the older men rarely take statutory samples, but when they do they tend more frequently to be followed by prosecution.
5. This is not to suggest that the use of the stat in such circumstances always leads to prosecution, but that such officers would prefer all of the (few) stats they take to be prosecuted.
6. Fault concepts have also been observed in the work of factory inspectors (Carson, 1970a; 1970b; 1980b; Law Commission, 1969) housing inspectors (Robert A. Kagan, private communication), public housing managers (Lempert, 1971), and child-support collection officials (Chambers, 1979).
7. Prior record or past history is an important source of context for judgment because it offers a longitudinal perspective; personal biography is profoundly significant in judging acts: see, e.g. Atkinson, 1978; Hawkins, 1981.
8. For example, arrest rates increase with antagonism to the police: Black, 1970; 1980; Black and Reiss, 1970; Piliavin and Briar, 1964; Preiss and Ehrlich, 1966; Sullivan and Siegel, 1972.
9. See n. 21, ch. 1.
10. Why this striking response was felt necessary is treated in ch. 10. This case, so far as I know, was the only 'blameless' pollution prosecuted in either agency in more than two years of fieldwork, and the only case prosecuted in the Crown Court.
11. Cf. Hawkins, 1981. An interesting feature arising from general discussion with agency staff is that when talking in the abstract about 'deliberate pollution', they assumed—in the absence of any information—serious consequences.
12. Assessments about the financial position of a polluter and his capacity to pay require some knowledge on the part of the field man of the costs of treatment plant and labour. Many claim not to know a great deal about these costs, though some possessed a fairly precise knowledge of certain items. Estimates of cost, if requested by the polluter, are normally couched, as a result, in rather vague terms.

Chapter 9: Using the Law

1. In talking of 'using the law' or referring to 'the formal process' in this chapter, I shall be concerned only with criminal prosecution. The water authorities in theory may also employ the civil law and apply for an injunction against a troublesome polluter. So far as I could discover, though the remedy is occasionally discussed and sometimes employed as a threat (see ch. 7), it has never been sought. In effect it means stopping a discharge, which would often result in a factory closure, a consequence generally recognized as too drastic to be realistically contemplated. (Injunction was, however, used quite frequently in the USA for breach of wartime regulations: Clinard, 1946.)
2. Other regulatory agencies make equally sparing use of the law. Cranston (1979:57), for example, discovered that one of his consumer protection departments mounted eleven prosecutions in six years, and even the department he described as having a higher rate of enforcement achieved only fifty-one in six years.
3. One of the most striking examples was of an officer in a rural catchment who had served over 20 years. He had been involved in only one prosecution, and that only peripherally. He had taken 20 or so formal samples in his career, which were all from one particularly bad silage season.
4. Probation officers are extremely reluctant to resort to the penal measure of imprisonment (Cicourel, 1968:229; Emerson, 1969:230). Cranston reports the same feeling of failure among his consumer protection officials (1979:99).
5. In deciding about prosecution, agencies are also concerned about affecting a victim's right to compensation. A conviction means that a pollution need not be proved in subsequent civil proceedings.
6. This feature is surprisingly neglected in those studies which have focused explicitly on decision making; two conspicuous exceptions, however, are Cicourel, 1968 and Emerson, 1969.
7. See also Cicourel, 1968. Ross (1970:62) observed the same practice among insurance claims adjusters.
8. See Silbey and Bittner, n.d.; Weaver, 1977:100. There is a similar preference to back only winning cases in criminal justice (Blumberg, 1967; Skolnick, 1967).
9. Cranston's study of prosecution files (1979:125) showed that in 'several' cases, 'legal proceedings had been abandoned simply because large organizations were prepared to call the bluff of the agency by making plain they would strenously contest the charge'.
10. A metaphor commonly used in this context by senior staff is that they are at the top of a mountain with a view on all sides into the far distance, whereas the field officer is at the bottom where his vision is limited.
11. Some field staff thought that the low sanction was itself good reason not to prosecute, a view shared by the nineteenth century factory inspectorate (Bartrip and Fenn, 1980b:96).
12. Polluters, it seems, rarely use lawyers. Only once in the fieldwork period did I observe negotiations in which a company's lawyer was present.
13. See also Mileski, 1971; Conklin, 1977. A variant of this view is that magistrates sympathize with firms faced with 'ever-increasing demands of officialdom' (Law Commission, 1969:51).
14. As it is for other enforcement agents. The traffic police studied by Gardiner who

were required to put in court appearances, for example, were less likely to write tickets: Gardiner, 1969:158.

Chapter 10: Law as Last Resort

1. Another senior man suggested his agency could increase its prosecution rate without difficulty 'because the more you do, the easier they get, so far as the evidence is concerned, 'cause the people know what they're doing'. Similarly, lack of resources is not thought a constraint in American antitrust enforcement. Weaver (1978:138) quotes an Assistant Attorney-General as saying 'I don't really know what we'd *do* with many more men.' (Emphasis in original)
2. Such evidence usually took the form of remarks of the sort quoted on p. 121.
3. An argument possibly adopted from the white-collar crime literature: see Newman, 1958; Sutherland, 1949.
4. This has also been observed in the enforcement of factory legislation: thus where 'arguments put to the court might show the [Factory] Inspectorate in an unfavourable light, for instance, in having failed in its duty of enforcement or in prosecuting an employer long after the required standards have been achieved, authorization of proceedings is likely to be withheld' (Law Commission, 1969:27).
5. Thus Banton (1964:136-7) has argued that informal controls (that is, a strategy of compliance) are more likely to be employed in areas of high social integration.
6. This is not to suggest, however, that the environmentalism which prompted the legislation is not a largely moralistic movement, or that strict liability does not express the view that those who pollute are morally obliged to be vigilant.
7. Newman (1957) showed that the public views adulteration of food, drugs, and cosmetics as more akin to traffic offences than to burglary. Perpetrators are 'lawbreakers' rather than 'criminals'.

Appendix

A Note on Research Methods

If studying the enforcement of regulation, why choose water pollution? There is much to be said when producing something of an exploratory analysis in relatively uncharted territory for selecting an arena of control that is bounded and manageable. Noise and air pollution may represent social problems equally worthy of enquiry, but as phenomena they are even more ephemeral than water pollution which at least possesses a kind of tangibility and largely predictable pathways of deviance.

A notion of the practical also informed the choice of the two regional water authorities studied, which were selected largely for their accessibility. Research was conducted in a pair of agencies on the assumption (subsequently shown to be unwarranted) that differences in both the character of the territory to be policed and organizational structures and procedures might lead to some substantial variation in enforcement behaviour. I have no reason to believe that the policies adopted by these authorities and the practices of their staffs are in any sense atypical of pollution control work in general; indeed, the parallels in the behaviour of staff in both agencies at all levels in the organizational hierarchy (and in the behaviour of other enforcement staff, as suggested in the literature) are striking.

The research field work reported in this study was carried out in agencies which had only been in existence in their present form little more than two years. Some legislation was awaiting implementation, and they were still actively engaged in developing and refining their administrative practices while I was working with them. Where it is relevant to do so I have discussed the nature and implications of such changes; otherwise I have, at the risk of some slight imprecision, used the present tense, frozen the action, and presented a picture of relatively stable organizational arrangements as they obtained in the mid to late 1970s. Doubtless there have been further changes since the field work was conducted, but this is not relevant, however, to the questions addressed in the book.

In carrying out the research, I was in contact with over seventy agency staff in the two regional water authorities for a period, with interruptions, of about thirty months. I worked with people at all levels in the organizational hierarchy, from director level to the newest and least experienced recruits. Though my research concerns demanded that I spent most of the time in the field with the pollution control officers, I also talked at considerable length with technical, administrative, and supervisory staff, and people employed in the agencies' legal departments.

Many social researchers who discuss the question of access to a research setting seem hardly able to conceal their surprise at the degree of co-operation they received and the apparent openness of their subjects. My experience was the same. I presented myself to senior staff when negotiating for access (and subsequently to field staff when the work was under way) as a researcher from Oxford University conducting a study concerned with the way pollution control worked and the part played in it by the law. I emphasized that I was interested in learning how field staff did their job, for which I would need to accompany them during their daily work. Senior staff were perfectly willing to co-operate. For my part I offered the usual assurances of anonymity and confidentiality, together with a promise that agency staff would be allowed to read a draft of the monograph, and have the opportunity of correcting errors and reflecting on the accuracy of the analysis. (In the event, there were few comments or corrections.)

Data were primarily collected by extensive participant observation of field staff in their routine work. The data from naturalistic observation were later supplemented by lengthy conversations in the field officers' area offices which were tape recorded. I also tape recorded long conversations with senior staff, including those in the legal departments, at headquarters. These conversations were rarely less than an hour in length, and usually much longer. I also made use of published agency materials and those internal agency documents to which I was given access: notes, memoranda, minutes of meetings, reports, and statements of policy.

The cases described in the book are those which simply happened to be dealt with at the time by field staff. In no sense did I either select a sample of cases, or deliberately follow through cases from beginning to end (though I did collect some data of this kind). Besides, with persistent failures to comply it is difficult to speak of the 'beginning' or 'end' of a case, when compliance is so negotiable. Some cases were observed in which a field man was embarking upon what he considered to be a potentially lengthy compliance process. In others, a particular 'problem' was already months or years old.

The field work was conducted in two stages, beginning in June 1976. The first was purely exploratory and partly intended to prepare the ground for more intensive work. I visited all the administrative areas in the northern authority in turn to talk to the various area supervisors and spent two further working days in the field with one of the pollution control officers. I also paid visits to two of the five administrative areas in the southern authority, spending time with almost all the field staff in both areas in the first phase of the research. (Owing to differences in internal organization the two areas selected for study in the southern authority were rather larger than the corresponding areas in the northern agency.) At this stage I also talked to the area supervisors in two other areas in the southern authority. The intention of this first phase of the research was to learn what enforcement work looked like, how field men defined problems, and what they regarded as their central concerns in the practical business of pollution control.

The second stage of the research began after an intermission of about three months devoted to preliminary analysis and reflection. Owing to other commitments it sprawled over a period of several months. I again sampled administrative units, not individual officers. I chose two of the seven administrative areas in the northern authority for a detailed study, working with all their field officers. These two areas were selected because they were recognized in the agency as containing more 'problems' than any of the other regions, thus promising a greater degree of enforcement activity. Since both areas had districts which ranged from the extremely rural to the entirely urban, there seemed to be little risk of producing a biased picture of the nature of enforcement. I also returned for further research to the two areas in the southern authority in which I had already worked. Both of these also contained a wide selection of problems characteristically associated with urban-industrial and rural life. It was important to observe field men handle this range and variety, since the tasks and tolerances involved in preserving the purity of a rural stream to permit potable supply or trout farming differ substantially from those to be confronted in attempting to prevent a highly polluted fishless urban stream from becoming a health hazard.

Though I hoped to be able to work with all the officers in the two selected southern authority areas, as I had in the northern, there were two officers with whom I did not make the rounds. In both cases their supervisor felt that they were not especially good at their job and was concerned (mistakenly, I thought) that I 'wouldn't learn anything from them'. In the interests of maintaining good relationships with the supervisor and his office in general, I did not demur.

I did the rounds with thirty-six field officers in this second phase, observing them at work throughout the day, usually for periods of two to four days. In total, I collected data from seventy-four different members of the pollution control divisions of the authorities. It was in this second stage that I concentrated particularly on the central research questions outlined in Chapter 1.

As other field workers have discovered, a lot is to be learned from travelling around with research subjects and hanging about in their offices. In my case I spent a good part of each day in the field men's cars, talking to them at length on the road between visits or at sampling stops. The pub at lunchtime, and sometimes in the evening as well, was also a place in which the field men would relax and reflect at length on their activities. I spent the whole of the time during fieldwork watching, listening, and talking. Where and whenever I could I scribbled brief notes which I later amplified and wrote up.

In doing the research none of the field staff raised difficulties or acted in any way which I could interpret as signifying concern about my presence. When out with them I sought to disguise my identity from no-one but left it to each field officer to decide whether he wanted to introduce me to dischargers. Some did; most did not. Sometimes I was introduced and identified as a researcher studying pollution control work; sometimes I was introduced neutrally by name alone but with no suggestion of my affiliation or why I was present. The evidence of subsequent conversations suggests that such introductions usually led the other party to assume that I was another member of the water authority. Only twice was I asked to identify myself, and I was never refused access to premises or land or excluded from negotiations.

I chose participant observation as the primary means of collecting data in observance of the interactionist injunction to respect and reflect the nature of the empirical world (Blumer, 1969). Observation provides the raw material which permits the activities of enforcement agents and their discretionary behaviour to be understood in the context of their routine work. It allows the researcher 'to attain a grasp of the meaning of ... rules as common-sense constructs from the perspective of those ... who promulgate and live with them' (Bittner, 1965:251). My chief concern, then, was to learn in detail at first hand about the business of pollution control at field level, to experience personally the field man's world, the mundane activities as well as the occasional dramatic events. I was careful not to interrupt the officers' daily routine and I tried to impress upon them that I wanted them to carry out their sampling, inspections, and negotiations exactly as they would in my absence. So far as I could tell, field staff generally observed this request, but there were a few occasions when I was given something of a guided tour by those officers who were keen to show me pollutions or special problems. Now and then—inevitably—I was taken to visit the sites of some of their more conspicuous successes.

Naturalistic observation was devoted, then, to learning what officers define as relevant in their everyday environment of sampling, monitoring, and routine encounters with dischargers and polluters. I did not study any of the pollution legislation before embarking on field work because I wanted to learn the law as the field officers knew it, and I wanted also to avoid the distortion arising from a particular sense of relevance which thorough prior knowledge of the formal structure of rules may have conferred on what I was actually observing. The perspective—the 'set of actions used by a group in solving collective problems' (Becker and Geer, 1960:280)—firmly adopted in this study is that of the pollution control officer. It is particularly important to bear this in mind in reading Part III of the book which analyses negotiating and bargaining tactics from one side only of a dual relationship (for a view of enforcement from the discharger's perspective, see Brittan's forthcoming study). The picture of the polluter which emerges is that as conceived by the field man. But it is this, of course, (as W. I. Thomas would doubtless have observed) upon which the officer premisses his enforcement strategy. Thorough immersion in the life of the pollution control officer makes it possible to draw inferences about the processes by which he defines and interprets the world. In an effort to determine the validity of the research interpretation these inferences were later tested by informal discussion with field staff to reach some conclusion about the extent to which they were recognizable to them (cf. Bittner, 1967a:701; Muir, 1977:287; see also Manning, 1972).

At the end of the fieldwork period, by which time very good relationships had been established with field staff, I tape-recorded conversations with almost all the officers I had observed at work during the second stage of the research and with almost all of their seniors. The purpose of the conversations was to discuss in a relatively systematic fashion some of the key issues which had emerged from participant observation. Extracts from them are presented in the text verbatim to preserve the sense of conversation, save that I have edited asides and the conversational bits and pieces, the 'ums', 'ahs', and 'ers', which litter naturally occurring talk. I use the conventions of points (...) to signify elided material, and a long dash (——) to indicate pauses. The discussions were very loosely structured to preserve as natural and informal a setting as possible and to avoid even the hint of a contrived formal interview (Schatzman and Strauss, 1973:71 ff.). In conducting the conversations at the end of the fieldwork period, when good rapport existed, my hope was that respondents would be less likely to create data, deliberately or unwittingly, either through a desire to mislead or faulty recall. I wished to avoid as much as possible the trap of making assumptions about behaviour from verbal descriptions (Phillips, 1973), hence the emphasis given to extensive prior participant observation.

Agency staff co-operated fully in the conversations without exception. All officials, field and headquarters alike, were (so far as I could tell) very open; some of them, certainly, were very frank. There were no objections to my using a tape recorder (indeed I was on two occasions actually offered the use of an office machine when I had technical problems with my own!). No meetings or encounters which I knew about were closed to me, and there were no signs that things were hidden from me or that staff acted in any sense evasively.

Data collection, of course, is a social process in itself and the presence of a researcher may well be a source of influence on the behaviour and responses of the research subjects. My impression is that my presence—if it had any impact at all—tended to affect only the trivial things in the daily round. For example, an officer who wanted to relieve the tedium of a routine (but time-consuming) river sample once observed 'Because you're here, I'd better take the sample properly.'

In presenting what I hope is an accurate picture of pollution control work I have made three closely-related assumptions. First, the longer the exposure of the researcher to his research subjects, the more familiar they will be with his presence; and the greater the exposure, the less their desire or ability to mislead. People do not keep up an act for long—and the job of being a field officer is more pressing than the task of managing a front towards an observer (Becker, 1970:46). The researcher's presence soon comes to be taken for granted (cf. Sudnow, 1967:7). This was a major reason for the relatively lengthy period of fieldwork. Participant observation, as Skolnick (1966:36) has observed:

> offers the subject less opportunity to dissimulate than he would have in answering a questionnaire, even if he were consciously telling the truth in response to standardized questions. ... The process of 'arguing', discussing, especially in the setting of the police work itself, creates an air of informality when opinions seem to be more openly expressed.

Secondly, research subjects are more likely to behave naturally—as they would, that is, in the absence of the observer—if the observer becomes a normal part of the scene. It does not mean that the researcher lurks unobtrusively in the background pretending not to be there (which is hardly possible when the observer is being ferried around by his research subjects), but rather that he helps out where need be, assisting with sampling, carrying equipment, and in general behaving as a colleague of the subject's might behave (cf. Schwartz and Schwartz, 1955; Sudnow, 1967: 7-8). There comes a point, of course, when playing the role of researcher as participant creates an ethical problem about the degree and kind of help which should be given. In my case, I decided to assist only in banal matters of fetching and carrying, and to withdraw when the help sought began to encroach on areas of judgment that could have a bearing on the handling of a case. (This rarely happened, in fact. The most dramatic example of an attempt to co-opt the researcher in solving problems arose when a field man of many years' experience was confronted with a massive oil pollution which had occurred through the failure of a new storage tank. The officer, however, could not attribute blame for the pollution and was torn between recommending prosecution for general deterrent purposes (given the size of the spill), and doing nothing more than issuing a stern warning. After presenting the facts of the case to me in great detail, he than asked 'What would you do?' in a tone of voice which indicated the question was not meant hypothetically. My response was entirely evasive (and no doubt terribly unhelpful to the officer concerned) since it merely rehearsed in somewhat different form the dilemma facing him.)

Finally, I take it as an indicator that people were behaving more or less normally that they were willing to talk freely about sensitive topics. If there was a field man who was thought to be lazy or no good at his job, for instance, others were prepared to tell me. If they had complaints about the agency or their superiors, as many did, again they were prepared to tell me. In both cases I assume that if field men were willing to talk critically—abusively, sometimes—about other staff, especially their senior staff, with whom (as they well knew) I was in personal contact, this is some indication of trust and rapport. In general, I take openness on the part of a research subject to be a favourable sign, whether the openness be of talk—about private ambition, for example, or bitterness towards senior staff—or of behaviour—such as an unhesitating willingness to engage in negotiations in the researcher's presence which could prove personally embarrassing.

Statutes

Control of Pollution Act 1974
Criminal Law Act 1977
Rivers (Prevention of Pollution) Act 1951
Rivers (Prevention of Pollution) Act 1961
Water Act 1973

Cases

Alphacell Ltd. v. Woodward [1972]2 A11ER 475 (HL); [1972] AC 824
Sherras v. De Rutzen (1895) 1 QB 918

References

Ackerman, Bruce A., Susan Rose Ackerman, James W. Sawyer jun., and Dale W. Henderson (1974) *The uncertain search for environmental quality* (New York: Free Press).

Adams, Richard N. and Jack J. Preiss (1960) *Human organization research: field relations and techniques* (Homewood, Ill.: Dorsey Press).

Allen, Francis A. (1977) 'Criminal law and the modern consciousness: some observations on blameworthiness' *Tennessee Law Review* 44: 735-63.

Anderson, James E. (1966) 'Public economic policy and the problem of compliance: notes for research' *Houston Law Review* 4: 62-72.

Arnold, Thurman W. (1935) *The symbols of government* (New Haven, Conn.: Yale University Press, repr. Harcourt Brace and World, 1962).

Atkinson, J. Maxwell (1978) *Discovering suicide. Studies in the social organization of sudden death* (Pittsburgh: University of Pittsburgh Press).

Aubert, Vilhelm (1952) 'White collar crime and social structure' *American Journal of Sociology* 58: 263-71.

Ball, Harry V. and Lawrence M. Friedman (1965) 'The use of criminal sanctions in the enforcement of economic legislation: a sociological view' *Stanford Law Review* 17: 197-223.

Banton, Michael (1964) *The policeman in the community* (London: Tavistock).

Barrett, J. W. (1977) 'We're good boys now—but can we stay in business?' *Product Finishing* March.

—— (1979) 'Prosecution may soon make cowboys of us all' *Product Finishing* 32: 12-14.

Bartrip, Peter W. J. (1979) 'Safety at work: the Factory Inspectorate in the fencing controversy, 1833-1857' Centre for Socio-Legal Studies, Working Paper no. 4.

Bartrip, Peter W. J., and P. T. Fenn (1980a) 'The conventionalization of factory crime—a re-assessment' *International Journal of the Sociology of Law* 8: 175-86.

—— (1980b) 'The administration of safety: the enforcement policy of the early Factory Inspectorate, 1844-1864' *Public Administration* 58: 87-102.

Bayley, David H. and Harold Mendelsohn (1968) *Minorities and the police. Confrontation in America* (New York: Free Press).

Beaumont, P. B. (1979) 'The limits of inspection: a study of the workings of the Government Wages Inspectorate' *Public Administration* 57: 203-17.

Becker, Howard S. (1963) *Outsiders. Studies in the sociology of deviance* (New York: Free Press).

—— (ed.) (1964) *The other side. Perspectives on deviance* (New York: Free Press).

—— (1970) *Sociological work. Method and substance* (Chicago: Aldine).

Becker, Howard S. and Blanche Geer (1960), 'Participant observation: the analysis of qualitative field data' in Adams and Preiss (1960): 267-89.

Bernstein, Marver H. (1955) *Regulating business by independent commission* (Princeton, NJ: Princeton University Press).

Bittner, Egon (1965) 'The concept of organization' *Social Research* 32: 239-55.

—— (1967a) 'The police on skid row: a study of peace keeping' *American Sociological Review* 32: 699-715.

—— (1967b) 'Police discretion in emergency apprehension of mentally ill persons' *Social Problems* 14: 278-92.

—— (1970) *The functions of the police in modern society* (Washington DC: National Institute of Mental Health).

—— (1974) 'Florence Nightingale in pursuit of Willie Sutton: a theory of the police' in Jacob (1974): 17-44.

Black, Donald J. (1970) 'Production of crime rates', *American Sociological Review* 35: 733-48.

—— (1971) 'The social organization of arrest' *Stanford Law Review* 23: 1087-111.

—— (1973) 'The mobilization of law' *Journal of Legal Studies* 2: 125-49.

—— (1976) *The behavior of law* (New York: Academic Press).

—— (1980) *The manners and customs of the police* (New York: Academic Press).

Black, Donald J. and Albert J. Reiss jun. (1967) 'Patterns of behavior in police and citizen transactions' in Field Surveys III: *Studies of Crime and Law Enforcement in Major Metropolitan Areas* ii. 1-139 for US President's Commission on Law Enforcement and Administration of Justice (Washington DC: US Government Printing Office).

—— (1970) 'Police control of juveniles' *American Sociological Review* 35: 63-77.

Blau, Peter M. (1963) *The dynamics of bureaucracy. A study of interpersonal relations in two government agencies* (Chicago: University of Chicago press, rev. ed.).

Blumberg, Abraham (1967) *Criminal justice* (Chicago: Quadrangle).

Blumer, Herbert (1969) *Symbolic interactionism: perspective and method* (Englewood Cliffs, NJ: Prentice-Hall).

Blumrosen, Alfred W. (1965) 'Antidiscrimination laws in action in New Jersey: a law-sociology study' *Rutgers Law Review* 19: 187-287.

Bordua, David J. (ed.) (1967) *The police: six sociological essays* (New York: Wiley).

Bordua, David J. and Albert J. Reiss jun. (1966) 'Command, control, and charisma: reflections on police bureaucracy' *American Journal of Sociology* 72: 68-76.

Brenner, Joel Franklin (1974) 'Nuisance law and the industrial revolution' *Journal of Legal Studies* 3: 403-33.

Brittan, Yvonne (forthcoming) *The impact of water pollution control on industry: a case study of fifty dischargers* (SSRC Centre for Socio-Legal Studies, Oxford).

Brown, M.A. (1976) 'Enforcement of oil pollution legislation. A practitioner's view' *Modern Law Review* 39: 162-8.

Burgess, Ernest W. (1950) 'Comment' and 'Concluding comment' to Hartung, *American Journal of Sociology* 56: 32-4.

Cain, Maureen (1973) *Society and the policeman's role* (London: Routledge & Kegan Paul).

Carlton, Richard E., Richard Landfield, and James B. Loken (1965) 'Enforcement of municipal housing codes' *Harvard Law Review* 78: 801-60.

Carson, W. G. (1970a) 'Some sociological aspects of strict liability and the enforcement of factory legislation' *Modern Law Review* 33: 396-412.

—— (1970b) 'White-collar crime and the enforcement of factory legislation' *British Journal of Criminology* 10: 383-98.

—— (1979) 'The conventionalization of early factory crime' *International Journal of the Sociology of Law* 7: 37-60.

—— (1980a) 'Early factory inspectors and the viable class society—a rejoinder' *International Journal of the Sociology of Law* 8: 187-91.

—— (1980b) 'The institutionalization of ambiguity: early British Factory Acts' in Geis and Stotland (1980): 142-73.

Chambers, David (1979) *Making fathers pay* (Chicago: University of Chicago Press).

Chatterton, Michael R. (1979) 'The supervision of patrol work under the fixed points system' in Holdaway (1979): 83-101.

Chevigny, Paul (1969) *Police power: police abuses in New York City* (New York: Vintage).

Cicourel, Aaron V. (1968) *The social organization of juvenile justice* (New York: Wiley).

Clinard, Marshall B. (1946) 'Criminological theories of violations of wartime regulations' *American Sociological Review* 11: 258-70.

—— (1952) *The black market. A study of white collar crime* (New York: Holt, Rinehart and Winston Inc., repr. Patterson Smith, NJ, 1969).

Clinard, Marshall B., and Richard Quinney (1967) *Criminal behavior systems: a typology* (New York: Holt, Rinehart and Winston).

Clinard, Marshall B., Peter C. Yeager, Jeanne Brissette, David Petrashek, and Elizabeth Harries (1979) *Illegal corporate behavior* (Washington DC: National Institute of Law Enforcement and Criminal Justice).

Conklin, John E. (1977) *'Illegal but not criminal'. Business crime in America* (Englewood Cliffs, NJ: Prentice-Hall).

Cranston, Ross (1979) *Regulating business. Law and consumer agencies* (London: Macmillan).

Crenson, Matthew A. (1971) *The un-politics of air pollution. A study of non-decisionmaking in the cities* (Baltimore: Johns Hopkins University Press).

Crozier, Michel (1964) *The bureaucratic phenomenon* (Chicago: University of Chicago Press).

Cumming, Elaine, Ian Cumming, and Laura Edell (1965) 'Policeman as philosopher, guide and friend' *Social Problems* 12: 276-86.

Currie, David P. (1975) 'Enforcement under the Illinois pollution law'

Northwestern University Law Review 70: 389-485.

Davies, J. Clarence (1970) *The politics of pollution* (Indianapolis: Bobbs-Merrill).

Dickens, Bernard M. (1970) 'Discretion in local authority prosecutions' *Criminal Law Review*: 618-33.

—— (1974) 'Law making and enforcement—a case study' *Modern Law Review* 37: 297-307.

Diver, Colin S. (1980) 'A theory of regulatory enforcement' *Public Policy* 28: 257-99.

Douglas, Jack D. (ed.) (1970) *Deviance and respectability. The social construction of moral meanings* (New York: Basic Books).

—— (ed.) (1972) *Research on deviance* (New York: Random House).

Edelhertz, Herbert and Thomas D. Overcast (eds.) (1980) *The development of a research agenda on white-collar crime* (Seattle: Battelle).

Edelman, Murray (1964) *The symbolic uses of politics* (Urbana, Ill.: University of Illinois Press).

Eisenberg, Melvin Aron (1976) 'Private ordering through negotiation: dispute-settlement and rulemaking' *Harvard Law Review* 89: 637-81.

Emerson, Robert M. (1969) *Judging delinquents. Context and process in juvenile court* (Chicago: Aldine).

Enthoven, Alain C. and A. Myrick Freeman III (eds.) (1973) *Pollution, resources, and the environment* (New York: Norton).

Ermann, M. David and Richard J. Lundman (eds.) (1978) *Corporate and governmental deviance. Problems of organizational behavior in contemporary society* (New York: Oxford University Press).

Esposito, John C. (1970) *Vanishing air*, The Ralph Nader Study Group Report on air pollution (New York: Grossman).

Fesler, James (1942) *The independence of state regulatory agencies* (Chicago: Public Administration Publication no. 85).

Frankel, Maurice (1974) *The Alkali Inspectorate. The control of industrial air pollution* (London: Social Audit).

Freedman, James O. (1975) 'Crisis and legitimacy in the administrative process' *Stanford Law Review* 27: 1041-76.

Freeman, A. Myrick III and Robert H. Haveman (1973) 'Clean rhetoric and dirty water' in Enthoven and Freeman (1973): 122-37.

Friedman, Lawrence M. (1967) 'Legal rules and the process of social change' *Stanford Law Review* 19: 786-840.

Fuller, Richard C. (1942) 'Morals and the criminal law' *Journal of Criminal Law, Criminology and Police Science* 32: 624-30.

Galanter, Marc (1974) 'Why the haves come out ahead: speculations on the limits of legal change' *Law and Society Review* 9: 95-160.

Gardiner, J. A. (1969) *Traffic and the police* (Cambridge, Mass.: Harvard University Press).

Geis, Gilbert (1967) 'White collar crime: the heavy electrical equipment antitrust cases of 1961' in Clinard and Quinney (1967): 139-51.

—— (1968) *White-collar criminal. The offender in business and the professions* (New York: Atherton).

Geis, Gilbert and Thomas R. Clay (1980) 'Criminal enforcement of California's Occupational Carcinogens Control Act' *Temple Law Quarterly* 53: 1067-99.

Geis, Gilbert and Robert W. Meier (eds.) (1977) *White collar crime* (New York: Free Press).

Geis, Gilbert and Ezra Stotland (eds.) (1980) *White-collar crime: theory and research* (Beverly Hills, Cal.: Sage).

Gellhorn, Ernest (1973) 'Adverse publicity by administrative agencies' *Harvard Law Review* 86: 1380-441.

Glaser, Daniel (ed.) (1974) *Handbook of criminology* (Chicago: Rand McNally).

Glenn, Michael K. (1973) 'The crime of pollution: the role of federal water pollution criminal sanctions' *American Criminal Law Review* 11: 835-82.

Goffman, Erving (1959) *The presentation of self in everyday life* (Garden City: Doubleday Anchor).

—— (1961) *Encounters. Two studies in the sociology of interaction* (Indianapolis: Bobbs-Merrill).

—— (1963) *Behavior in public places. Notes on the social organization of gatherings* (New York: Free Press).

—— (1967) *Interaction ritual. Essays on face-to-face behavior* (New York: Anchor Books).

—— (1970) *Strategic interaction* (Oxford: Blackwell).

—— (1971) *Relations in public. Microstudies of the public order* (Harmondsworth: Penguin).

Goldstein, H. (1963) 'Police discretion: the ideal versus the real' *Public Administration* 23: 140-8.

Goldstein, Paul and Robert Ford (1972) 'The management of air quality: legal structures and official behavior' *Buffalo Law Review* 21: 1-48.

Gouldner, Alvin W. (1954) *Patterns of industrial bureaucracy* (New York: Free Press).

—— (1960) 'The norm of reciprocity—a preliminary statement' *American Sociological Review* 25: 161-78.

Gribetz, Judah and Frank P. Grad 'Housing code enforcement: sanctions and remedies' *Columbia Law Review* 66: 1254-90.

Gross, Edward (1980) 'Organization structure, and organizational crime' in Geis and Stotland (1980): 52-76.

Guardian (1980) 'Cancer peril in the Rhine' August 6.

Gunningham, Neil (1974) *Pollution, social interest and the law* (London: Martin Robertson).

Gusfield, Joseph (1967) 'Moral passage: the symbolic process in public designations of deviance' *Social Problems* 15: 175-88.

Hagevik, George (1968) 'Legislation for air quality management: reducing theory to practice' *Law and Contemporary Problems* 33: 369-98.

Hartung, Frank E. (1950) 'White collar offences in the wholesale meat industry in Detroit' *American Journal of Sociology* 56: 25-32.

Harvard Law Review (1979) 'Developments in the law—corporate crime: regulating corporate behavior through criminal sanctions' *Harvard Law Review* 92: 1227-375.

Hawkins, Keith (1981) 'The interpretation of evil in criminal settings' in Ross (1981): 99-126.

Herring, Pendleton (1936) *Public administration and the public interest* (New York: McGraw Hill).

Hines, N. William (1966a) 'Nor any drop to drink: public regulation of water quality. Part I: State pollution control programs' *Iowa Law Review* 52: 186-235.

—— (1966b) 'Nor any drop to drink: public regulation of water quality. Part II: Interstate arrangements for pollution control' *Iowa Law Review* 52: 432-57.

Holdaway, Simon (ed.) (1979) *The British police* (London: Edward Arnold).

Holden, Matthew jun. (1966) 'Pollution control as a bargaining process: an essay on regulatory decision-making' (Ithaca, NY: Cornell University Water Resources Center).

Holmes, Oliver Wendell (1881) *The common law* (Boston: Little, Brown).

Hucke, Jochen (1978) 'Bargaining in regulative policy implementation: the case of environmental policy in the Federal Republic of Germany', paper prepared for Workshop on Inter-organizational Networks in Public Policy Implementation, European Consortium for Political Research, Grenoble, April.

Hughes, Everett C. (1971) *The sociological eye. Selected papers on work, self and the study of society* (Chicago: Aldine-Atherton).

Huntington, Samuel P. (1952) 'The marasmus of the ICC: the Commission, the railroads, and the public interest' *Yale Law Journal* 61: 467-509.

Irwin, William A. (1970) 'Michigan air pollution control: a case study' *Journal of Law Reform* 4: 23-46.

Jacob, Herbert (ed.) (1974) *The potential for reform of criminal justice* (Beverly Hills, Cal.: Sage).

Jacobs, Francis G. (1971) *Criminal responsibility* (London: Weidenfeld and Nicolson).

Johnson, John M. (1972) 'The practical uses of rules' in Scott and Douglas (1972): 215-48.

Joseph, Nathan and Nicholas Alex (1971) 'The uniform: a sociological perspective' *American Journal of Sociology* 77: 719-30.

Jowell, Jeffrey L. (1977a) 'Bargaining in development control' *Journal of Planning and Environmental Law:* 414-33.

—— (1977b) 'The limits of law in urban planning' *Current Legal Problems* 30: 63-83.

Kadish, Sanford H. (1963) 'Some observations on the use of criminal sanctions

in enforcing economic regulations' *University of Chicago Law Review* 30: 423-49.
Kagan, Robert A. (1978) *Regulatory justice: implementing a wage-price freeze* (New York: Russell Sage).
—— (1980) 'The positive uses of discretion: the good inspector', paper presented to the 1980 Annual Meeting of the Law and Society Association, Madison, Wisconsin, June.
Kagan, Robert A. and John T. Scholz (1979) 'The criminology of the corporation and regulatory enforcement strategies', paper presented to the Symposium on Organizational Factors in the Implementation of Law, University of Oldenburg, May.
Katona, George (1945) *Price control and business. Field studies among producers and distributors of consumer goods in the Chicago area, 1942-44* (Bloomington, Ind.: Principia Press).
Katzman, Robert A. (1980) *Regulatory bureaucracy. The Federal Trade Commission and the antitrust policy* (Cambridge, Mass.: MIT Press).
Kaufmann, Herbert (1960) *The forest ranger. A study in administrative behavior* (Baltimore: Johns Hopkins University Press).
Kelman, Steven (1981) *Regulating America, regulating Sweden: a comparative study of occupational safety and health policy* (Cambridge, Mass.: MIT Press).
Kneese, Allen V. (1973) 'Economics and the quality of the environment: some empirical experiences' in Enthoven and Freeman (1973): 72-87.
Kolko, Gabriel (1965) *Railroads and regulations* (Princeton, NJ: Princeton University Press).
—— (1967) *The triumph of conservatism. A reinterpretation of American history, 1900-1916* (Chicago: Quadrangle).
LaFave, Wayne R. (1962) 'The police and non-enforcement of the law' *Wisconsin Law Review*, Part I: 104-37; Part II: 179-239.
—— (1965) *Arrest: the decision to take a suspect into custody* (Boston: Little, Brown).
Landis, James M. (1938) *The administrative process* (New Haven, Conn.: Yale University Press).
Lane, Robert E. (1953) 'Why business men violate the law' *Journal of Criminal Law, Criminology and Police Science* 44: 151-65.
—— (1954) *The regulation of businessmen* (New Haven, Conn.: Yale University Press).
Law Commission (1969) *Strict liability and the enforcement of the Factories Act 1961*, a report by members of the Sub-Faculty of Law at the University of Kent at Canterbury to the Law Commission, December. Working Paper no. 30.
Leiserson, Avery (1942) *Administrative regulation: a study in representation of interests* (Chicago: University of Chicago Press).
Lemert, Edwin M. (1972) *Human deviance, social problems and social control* (2nd ed.) (Englewood Cliffs, NJ: Prentice-Hall).
Lempert, Richard O. (1971) 'Evictions from public housing: a sociological inquiry' Ph.D. dissertation, University of Michigan.

Lidstone, K. W., Russell Hogg, and Frank Sutcliffe, with A. E. Bottoms and Monica A. Walker (1980) *Prosecutions by private individuals and non-police agencies* Royal Commission on Criminal Procedure Research Study no. 10 (London: HMSO).

Lipsky, Michael (1976) 'Towards a theory of street-level bureaucracy', in *Theoretical perspectives on urban politics* (Englewood Cliffs, NJ: Prentice Hall)

Long, Susan B. (1979), 'The Internal Revenue Service: examining the exercise of discretion in tax enforcement', paper presented to 1979 Annual Meeting of the US Law and Society Association, San Francisco, May.

Lowi, Theodore J. (1979) *The end of liberalism. The second Republic of the United States* (2nd ed.) (New York: Norton).

Macaulay, Stewart (1963) 'Non-contractual relations in business: a preliminary study' *American Sociological Review* 28: 55-67.

McCabe, Sarah and Frank Sutcliffe (1978) *Defining crime: a study of police decisions* (Oxford: Blackwell).

McCleary, Richard (1978) *Dangerous men. The sociology of parole* (Beverly Hills, Cal.: Sage).

McHugh, Peter (1970) 'A common-sense conception of deviance' in Douglas (1970): 61-88.

McKie, James W. (ed.) (1975) *Social responsibility and the business predicament* (Washington DC: Brookings Institution).

McLoughlin, J. (1976) *The law and practice relating to pollution control in the United Kingdom* (London: Graham and Trotman).

Manning, Peter K. (1972) 'Observing the police: deviants, respectables, and the law' in Douglas (1972): 213-68.

—— (1977) *Police work: the social organization of policing* (Cambridge, Mass.: MIT Press).

—— (1980) *The narcs' game. Organizational and informational limits on drug law enforcement* (Cambridge, Mass: MIT Press).

March, James and Herbert Simon (1958) *Organizations* (New York: Wiley).

Matza, David (1964) *Delinquency and drift* (New York: Wiley).

Mawby, R. I. (1979) 'Policing by the Post Office' *British Journal of Criminology* 19: 242-53.

Mayhew, Leon H. (1968) *Law and equal opportunity. A study of the Massachusetts Commission against discrimination* (Cambridge, Mass.: Harvard University Press).

Mileski, Maureen (1971) 'Policing slum landlords: an observation study of administrative control' Ph.D. dissertation, Yale University.

Morris, Joe Scott (1972), 'Environmental problems and the use of criminal sanctions' *Land and Water Law Review* 7: 421-31.

Muir, William Ker jun. (1977) *Police. Streetcorner politicians* (Chicago: University of Chicago Press).

Murphy, Earl Finbar (1961) *Water purity. A study in legal control of natural resources* (Madison: University of Wisconsin Press).

Nader, Ralph (1970) Foreword to Esposito (1970).
Nagel, Stuart (1974) 'Incentives for compliance with environmental law' *American Behavioral Scientist* 17: 690-710.
Newman, Donald J. (1957) 'Public attitudes toward a form of white collar crime' *Social Problems* 4: 228-32.
—— (1958) 'White-collar crime' *Law and Contemporary Problems* 23: 735-53.
Nicholson, N. J. (1973) 'Water pollution' unpublished paper.
Niederhoffer, Arthur (1967) *Behind the shield. The police in urban society* (New York: Doubleday).
Nivola, Pietro S. (1978) 'Distributing a municipal service: a case study of housing inspection' *Journal of Politics* 59: 81.
—— (1979) *The urban service problem. A study of housing inspection* (Lexington, Mass.: Lexington Books).
Nonet, Philippe (1969) *Administrative justice. Advocacy and change in a government agency* (New York: Russell Sage).
Observer (1980) 'Clean water pledged for all', November 30.
Parnas, Raymond I (1967) 'The police response to the domestic disturbance' *Wisconsin Law Review*: 914-60.
Paulus, Ingeborg (1974) *The search for pure food. A sociology of legislation in Britain* (London: Martin Robertson).
Pepinsky, Harold E. (1976) *Crime and conflict. A study of law and society* (London: Martin Robertson).
Petersen, David M. (1968) 'The police, discretion and the decision to arrest' Ph.D. dissertation, University of Kentucky.
—— (1971) 'Informal norms and police practice: the traffic ticket quota system' *Sociology and Social Research* 55: 354-62.
Phillips, Derek L. (1973) *Knowledge from what? Theories and methods in social research* (Chicago: Rand McNally).
Piliavin, Irving and Scott Briar (1964) 'Police encounters with juveniles' *American Journal of Sociology* 70: 206-14.
Preiss, Jack J. and Howard J. Ehrlich (1966) *An examination of role theory: the case of the state police* (Lincoln, Neb.: University of Nebraska Press).
Prottas, Jeffrey Manditch (1979) *People processing. The street-level bureaucrat in public service bureaucracies* (Lexington, Mass.: Lexington Books).
Punch, Maurice and Trevor Naylor (1973) 'The police: a social service' *New Society* 24 no. 554: 358-61 (17 May).
Rabin, Robert L. (1972) 'Agency criminal referrals in the Federal System: an empirical study of prosecutorial discretion' *Stanford Law Review* 24: 1036-91.
Reiss, Albert J. jun. (1966) 'The study of deviant behavior: where the action is' *Ohio Valley Sociologist* 32.
—— (1971) *The police and the public* (New Haven, Conn.: Yale University Press).
—— (1974) 'Discretionary justice' in Glaser (1974): 679-99.
—— (1980) 'The policing of organizational life', paper prepared for

International Seminar on 'Management and Control of Police Organization' Nijenrode, Netherlands, December.

Reiss, Albert J. jun. and Albert D. Biderman (1980) *Data sources on white-collar law-breaking* (Washington DC: National Institute of Justice).

Reiss, Albert J. jun. and David J. Bordua (1967) 'Environment and organization: a perspective on the police' in Bordua (1967): 25-55.

Richardson, Genevra with A. I. Ogus and Paul Burrows (forthcoming) *Policing pollution: a study of regulation and enforcement* (Oxford: Clarendon Press).

Robison, James and Paul T. Takagi (1968) *Case decisions in a state parole system* (Sacramento, Cal.: California Department of Corrections) Research Report no. 31.

Rock, Paul (1968) 'Observations on debt collection', *British Journal of Sociology* 19: 176-90.

—— (1973a) *Deviant behaviour* (London: Hutchinson).

—— (1973b) *Making people pay* (London: Routledge & Kegan Paul).

Roos, Leslie L. and Noralou P. Roos (1972) 'Pollution, regulation, and evaluation' *Law and Society Review* 6: 509-29.

Ross, H. Laurence (1970) *Settled out of court. The social process of insurance claims adjustment* (Chicago: Aldine).

—— (ed.) (1981) *Law and deviance* (Beverly Hills, Cal.: Sage).

Ross, H. Laurence and John M. Thomas (1981) 'Blue-collar bureaucrats and the law in action: housing code regulation in three cities' unpublished paper.

Roth, Julius A. (1963) *Timetables: structuring the passage of time in hospital treatment and other careers* (Indianapolis: Bobbs-Merrill).

Rourke, Francis E. (1957) 'Law enforcement through publicity' *University of Chicago Law Review* 24: 225-55.

Royal Commission on Sewage Disposal (1912) Eighth Report: *Standards and tests for sewage and sewage effluents discharging into rivers and streams* Cd. 6464.

Rubin, Jesse (1972) 'Police identity and the police role' in Steadman (1972): 12-50.

Rubinstein, Jonathan (1973) *City police* (New York: Ballantine Books).

Sabatier, Paul (1975) 'Social movements and regulatory agencies: toward a more adequate—and less pessimistic—theory of clientele capture' *Policy Sciences* 6: 301-42.

Sacks, Harvey (1972) 'Notes on police assessment of moral character' in Sudnow (1972): 280-93.

Sanders, William B. (1977) *Detective work. A study of criminal investigations* (New York: Free Press).

Sayre, Francis Bowes (1933) 'Public welfare offences' *Columbia Law Review* 33: 55-88.

Schatzman, Leonard and Anselm L. Strauss (1973) *Field research. Strategies for a natural sociology* (Englewood Cliffs, NJ: Prentice-Hall).

Schrager, Laura Shill and James F. Short jun. (1978) 'Toward a sociology of organizational crime' *Social Problems* 25: 407-19.

—— (1980) 'How serious a crime? Perceptions of organizational and common crimes' in Geis and Stotland (1980): 14-31.

Schuck, Peter (1972) 'The curious case of the indicted meat inspectors' *Harper's Magazine* September: 81-8.

—— (1979) 'Litigation, bargaining and regulation' *Regulation* July/ August: 26-34

Schwartz, Morris S. and Charlotte Green Schwartz (1955) 'Problems in participant observation' *American Journal of Sociology* 60: 343-53.

Scott, Marvin B. and Stanford M. Lyman (1970) 'Accounts, deviance, and social order' in Douglas (1970): 89-119.

Scott, Robert A. and Jack D. Douglas (eds.) (1972) *Theoretical perspectives on deviance* (New York: Basic Books).

Selznick, Philip (1966) *TVA and the grass roots. A study in the sociology of formal organization* (rev. ed.) (New York: Harper and Row).

—— (1969) Foreword to Nonet (1969).

Sewell, W. R. Derrick and Lorna R. Barr (1977) 'Evolution in the British institutional framework for water management' *Natural Resources Journal* 17: 395-413.

Shapiro, Susan (1980) 'Thinking about white collar crime: matters of conceptualization and research' (Washington DC: National Institute of Justice).

Shibutani, Tamotsu (1961) *Society and personalty. An interactionist approach to social psychology* (Englewood Cliffs, NJ: Prentice-Hall).

Silbey, Susan S. (1978) 'Consumer justice: the Massachusetts Attorney General's Office of Consumer Protection, 1970-74' Ph.D. dissertation, University of Chicago.

—— (n.d.) 'Mediation: a means of cooperating with business' unpublished paper.

Silbey, Susan S. and Egon Bittner (n.d.) 'The availability of legal devices' unpublished paper.

Skolnick, Jerome H. (1966) *Justice without trial. Law enforcement in democratic society* (New York: Wiley).

—— (1967) 'Social control in the adversary system' *Journal of Conflict Resolution* 11: 52-70.

Skolnick, Jerome H. and J. Richard Woodworth (1967) 'Bureaucracy, information, and social control: a study of a morals detail' in Bordua (1967): 99-136.

Smith, Miles and Anthony Pearson (1969) 'The value of strict liability' *Criminal Law Review:* 5-16.

Staw, Barry M. and Eugene Szwajkowski (1975) 'The scarcity-munificence component of organizational environments and the commission of illegal acts' *Administrative Science Quarterly* 20: 345-54.

Steadman, Robert F. (ed.) (1972) *The police and the community* (Baltimore: Johns

Hopkins University Press).
Stearns, L. Reeve (1979) 'Fact and fiction of a model enforcement bureaucracy: the labour inspectorate of Sweden' *British Journal of Law and Society* 6: 1-23.
Steele, Eric H. (1975) 'Fraud, dispute, and the consumer: responding to consumer complaints' *University of Pennsylvania Law Review* 123: 1107-86.
Stinchcombe, Arthur L. (1963) 'Institutions of privacy in the determination of police administrative practice' *American Journal of Sociology* 69: 150-60.
Stjernquist, Per (1973) *Laws in the forests. A study of public direction of Swedish private forestry* (Lund: C.W.K. Gleerup).
Stone, Christopher D. (1975) *Where the law ends. The social control of corporate behavior* (New York: Harper and Row).
Strauss, Anselm (1978) *Negotiations. Varieties, contexts, processes, and social order* (San Francisco: Jossey-Bass).
Sudnow, David (1965) 'Normal crimes: sociological features of the penal code in a public defender office' *Social Problems* 12: 255-76.
—— (1967) *Passing on. The social organization of dying* (Englewood Cliffs, NJ: Prentice-Hall).
—— (ed.) (1972) *Studies in social interaction* (New York: Free Press).
Sullivan, Dennis C. and Larry J. Siegel (1972) 'How police use information to make decisions: an application of decision games' *Crime and Delinquency* 18: 253-62.
Sutherland, Edwin H. (1940) 'White collar criminality' *American Sociological Review* 5: 1-12.
—— (1945) 'Is white collar crime crime?' *American Sociological Review* 10: 132-39.
—— (1949) *White collar crime* (New York: The Dryden Press).
Sykes, Gresham M. and David Matza (1957) 'Techniques of neutralization: a theory of delinquency' *American Sociological Review* 22: 664-70.
Tappan, Paul W. (1947) 'Who is the criminal?' *American Sociological Review* 12: 96-102.
Texas Law Review (1970) 'Current problems: water pollution control in Texas' *Texas Law Review* 48: 1029-182.
Thomas, John M. (1980) 'The regulatory role in the containment of corporate illegality' in Edelhertz and Overcast (1980): 107-29.
Thompson, Victor A. (1950) *The regulatory process in OPA rationing* (New York: King's Crown Press).
Tiffany, Lawrence P., Donald M. McIntyre jun., and Daniel L. Rotenberg (1967) *Detection of crime. Stopping and questioning, search and seizure, encouragement and entrapment* (Boston: Little, Brown).
Times, The (1978) 'Creating the world's best super-structure' September 18.
Truman, David Bicknell (1940) *Administrative decentralization. A study of the Chicago field offices of the United States Department of Agriculture* (Chicago: University of Chicago Press).

Van Maanen, John (1973) 'Observations on the making of policemen' *Human Organization* 32: 407-18.

—— (1974) 'Working the street: a developmental view of police behavior' in Jacob (1974): 83-130.

Walsh, James Leo (1972) 'Research note: Cops and stool pigeons—professional striving and discretionary justice in European police work' *Law and Society Review* 7: 299-306.

Walsh, Marilyn E. and Donna D. Schram (1980) 'The victim of white-collar crime: accuser or accused?' in Geis and Stotland (1980): 32-51.

Washington Post (1980) 'Water rescue' March 12.

—— (1981) 'U.S. relaxing enforcement of regulations' November 15.

Weaver, Suzanne (1977) *Decision to prosecute: organization and public policy in the Antitrust Division* (Cambridge, Mass.: MIT Press).

Weber, Max (1946) *From Max Weber: essays in sociology* translated by H. H. Gerth and C. Wright Mills (New York: Oxford University Press).

—— (1947) *The theory of social and economic organization* translated by A. M. Henderson and Talcott Parsons (New York: Oxford University Press).

—— (1958) *The protestant ethic and the spirit of capitalism* translated by Talcott Parsons (New York: Charles Scribner's Sons).

Wenner, Lettie M. (1972) 'Enforcement of water pollution control law' *Law and Society Review* 6: 481-507.

Werthman, Carl and Irving Piliavin (1967) 'Gang members and the police' in Bordua (1967): 56-98.

Westley, William S. (1970) *Violence and the police: a sociological study of law, custom, and morality* (Cambridge, Mass.: MIT Press).

Wheeler, Stanton (ed.) (1968a) *Controlling delinquents* (New York: Wiley).

Wheeler, Stanton, Edna Bonacich, M. Richard Cramer, and Irving K. Zola (1968b) 'Agencies of delinquency control: a comparative analysis' in Wheeler (1968a): 31-60.

Whyte, William Foote (1943) *Street-corner society. The social structure of an Italian slum* (Chicago: University of Chicago Press).

Wilson, James Q. (1963) 'The police and their problems: a theory' *Public Policy* 12: 189-216.

—— (1968) *Varieties of police behavior. The management of law and order in eight communities* (Cambridge, Mass.: Harvard University Press).

—— (1975) 'The politics of regulation' in McKie (1975).

—— (1978) *The investigators. Managing FBI and Narcotics Agents* (New York: Basic Books).

—— (1980a) 'The politics of regulation' in Wilson (1980b): 357-94.

—— (ed.) (1980b) *The politics of regulation* (New York: Basic Books).

Yoder, Stephen A. (1978) 'Criminal sanctions for corporate illegality' *Journal of Criminal Law and Criminology* 69: 40-58.

Zwick, David and Marcy Benstock (1971) *Water wasteland* (New York: Grossman).

Author Index

Ackerman, B.A., 10, 25, 212
Alex, N., 134
Allen, F.A., 14
Anderson, J.E., 209, 219
Arnold, T.W., 194
Atkinson, J.M., 221
Aubert, V., 210

Ball, H.V., 10
Banton, M., 209, 211, 213, 223
Barr, L.R., 211
Barrett, J.W., 112, 219
Bartrip, P.W.J., 210, 222
Bayley, D.H., 211
Beaumont, P.B., 209
Becker, H.S., 12, 15, 23, 72, 210, 229, 231
Benstock, M., 3
Bernstein, M.H., 3
Biderman, A.D., 5, 6, 14, 209, 211, 217
Bittner, E., 57, 108, 110, 127, 147, 196, 209, 211, 219, 222, 229
Black, D.J., 3, 4, 97, 211, 216, 217, 221
Blau, P.M., 17, 54, 61, 121, 122, 204, 213
Blumberg, A., 190, 222
Blumer, H., 228
Blumrosen, A.W., 209
Bordua, D.J., 8, 66, 211, 216, 217
Brenner, J.F., 11
Briar, S., 211, 220, 221
Brittan, Y., 120, 143, 150, 151, 187, 207, 219, 229
Brown, M.A., 221
Burgess, E.W., 210

Cain, M., 7, 60, 211
Carlton, R.E., 209
Carson, W.G., 11, 204, 209, 210, 213, 220, 221
Chambers, D., 221
Chatterton, M.R., 209, 213, 216
Chevigny, P., 211
Cicourel, A.V., 15, 67, 222
Clay, T., 209
Clinard, M.B., 17, 191, 203, 209, 210, 222
Conklin, J.E., 191, 222
Cranston, R., 81, 94, 97, 131, 188, 191, 209, 212, 213, 219, 222
Crenson, M.A., 213
Crozier, M., 54
Cumming, E., 209
Currie, D.P., 220

Davies, J.C., 192, 212
Dickens, B.M., 117, 209, 221
Dilhorne, Viscount, 211
Diver, C.S., 11, 195, 210

Edelhertz, H., 211
Edelman, M., xii, 9, 118, 122
Ehrlich, H.J., 211, 221
Eisenberg, M.A., 5, 122
Emerson, R.M., 15, 52, 98, 119, 162, 165, 220, 222
Ermann, M.D., 211

Fenn, P.T., 210, 222
Fesler, J., 209
Ford, R., 110, 209
Frankel, M., 209
Freedman, A.M., 3
Friedman, L.M., 10, 12
Fuller, R.C., 10

Galanter, M., 43, 120
Gardiner, J.A., 196, 220, 222, 223
Geer, B., 229
Geis, G., 17, 209, 210
Gellhorn, E., 219
Glenn, M.K., 209
Goffman, E., 15, 42, 77, 134, 171, 220
Goldstein, H., 211
Goldstein, P., 110, 209
Gouldner, A.W., 123, 216, 220
Grad, F.P., 209
Gribetz, J., 209
Guardian, The, xi
Gunningham, N., 3, 209
Gusfield, J., 194

Hagevik, G., 220

Hartung, F.E., 210
Harvard Law Review, 13
Haveman, R.H., 3
Hawkins, K.O., 221
Herring, P., 209
Hines, N.W., 209
Holden, M., jun., 110, 209, 220
Holmes, O.W., 203
Hucke, J., 123
Hughes, E.C., 47
Huntington, S.P., 209

Irwin, W.A., 209

Jacobs, F.G., 13
Johnson, J.M., 216
Joseph, N., 134
Jowell, J.L., 220

Kadish, S.H., 10, 110
Kagan, R.A., 9, 112, 115, 181, 200, 209, 212, 218, 219, 221
Katona, G., 110
Katzman, R.A., 209, 217
Kaufman, H., 16, 28, 70, 124, 214, 216
Kelman, S., 209, 219
Kneese, A.V., 24
Kolko, G., 210

LaFave, W.R., 209, 211, 217, 218, 220
Landis, J.M., 148
Lane, R.E., 110, 203, 219
Law Commission, 209, 210, 220, 221, 222, 223
Leiserson, A., 209
Lemert, E.M., 10, 13, 166
Lempert, R.O., 221
Lidstone, K.W., 209
Lipsky, M., 57
Long, S.B., 90, 192, 210
Lowi, T.J., 11, 22
Lundman, R.J., 211
Lyman, S.M., 171

Macaulay, S., 198
Manning, P.K., 13, 15, 16, 45, 61, 67, 153, 194, 195, 201, 210, 211, 213, 214, 217, 229
March, J.G., 54
Matza, D., 10, 118, 163, 191, 203, 210, 213
Mawby, R.I., 81, 213
Mayhew, L.H., 209
McCabe, S., 210
McCleary, R., 68
McHugh, P., 163
McLoughlin, J., 211
Meier, R.W., 210
Mendelsohn, H., 211
Mileski, M., 6, 81, 192, 209, 222
Morris, J.S., 210
Muir, W.K. jun., 154, 211, 220, 229
Murphy, E.F., 9, 210

Nader, R., 192
Nagel, S., 126
Naylor, T., 210
Newman, D.J., 223
Nicholson, N.J., 115
Niederhoffer, A., 66, 211
Nivola, P.S., 209

Observer, The, xi
Overcast, T.D., 211

Parnas, R.I., 210
Paulus, I, 12, 13, 189, 209, 210
Pearson, A., 209
Pepinsky, H.E., 144
Petersen, D.M., 67, 196, 211, 214, 217, 218
Phillips, D.L., 230
Piliavin, I., 84, 211, 220, 221
Preiss, J.J., 211, 221
Prottas, J.M., 54, 57
Punch, M., 210

Rabin, R.L., 201
Reiss, A.J. jun., 5, 6, 8, 14, 15, 66, 90, 144, 209, 211, 216, 217, 221
Richardson, G., 211, 212, 213
Robison, J.O., 181
Rock, P., 77, 129, 209, 218, 220
Roos, L.L., 218
Roos, N.P., 218
Ross, H.L., 15, 41, 58, 118, 209, 210, 213, 216, 217, 218, 220, 221, 222
Roth, J.A., 218
Rourke, F.E., 219
Rubin, J., 95
Rubinstein, J., 46, 66, 67, 83, 84, 173, 211, 216, 217

Sabatier, P., 209
Sacks, H., 77, 217
Salmon, Lord, 211
Sanders, W.B., 89
Sayre, F.B., 13
Schatzman, L., 230
Scholz, J.T., 112, 200, 219

Schrager, L.S., 12, 211
Schram, D.D., 11
Schuck, P., 122, 144, 209, 216
Schwartz, C.G., 231
Schwartz, M.S., 231
Scott, M.B., 171
Selznick, P., xiv, 3, 193, 195
Sewell, W.R.D., 211
Shapiro, S., 211
Shibutani, T., 53
Short, J.F. jun., 12, 211
Siegel, L.J., 221
Silbey, S.S., 67, 111, 127, 147, 209, 220, 222
Simon, H.S., 54
Skolnick, J.H., 15, 58, 66, 67, 84, 98, 188, 211, 214, 216, 222, 231
Smith, M., 209
Staw, B.M., 110
Stearns, L.R., 209
Steele, E.H., 209
Stinchcombe, A.L., 142
Stjernquist, P., 81, 115, 134, 209, 213, 215, 220
Stone, C.D., 203
Strauss, A.L., 220, 230
Sudnow, D., 15, 52, 162, 231
Sullivan, D.C., 221
Sutcliffe, F., 210
Sutherland, E.H., 11, 17, 210, 223
Sykes, G.M., 118
Szwajkowski, E., 110

Takagi, P.T., 181
Tappan, P.W., 210
Texas Law Review, 209
Thomas, J.M., 11, 209, 210, 211, 218, 220
Thomas, W.I., 229
Thompson, V.A., 124, 209
Tiffany, L.P., 218
Times, The, 18
Truman, D.B., 209

Van Maanen, J., 46, 49, 134, 216

Walsh, J.L., 98
Walsh, M.E., 11
Washington Post, xi, 209
Weaver, S., 181, 217, 218, 222, 223
Weber, M., 213
Wenner, L.M., 67
Werthman, C., 84, 220
Westley, W.S., 211, 220
Wheeler, S., 189
Whyte, W.F., 196
Wilberforce, Lord, 211
Wilson, J.Q., 49, 54, 57, 59, 69, 98, 196, 209, 210, 211, 216, 219
Woodworth, J.R., 214
Wright, Mr Justice, 211

Yoder, S.A., 10

Zwick, D., 3

Subject Index

Alphacell v *Woodward*, 211 n.29
ambivalence, moral and political, 8-13, 193, 195, 203-7
area supervisors, 51, 61-3, 216 n.5
authority, presentation of, 133-41

bargaining, 122-8, 200
blameworthiness, 161-71, 196, 202-7
bluffing, 149-54, 160, 171-2

competence of field staff, 66-70
complaints about pollution, 90-1, 96-8
compliance, 15
  delay in, 119-20
  nature of, 105ff, 146, 182, 197
compliance systems or strategies, 3-8, 105ff, 122-8, 129ff, 134, 153-4, 173-4, 178-9, 190, 195-8, 202
consents, *see* standards
control, extension of, 147-54
Control of Pollution Act 1974
  implications for enforcement, 33-5
co-opting of agency staff, 31-2, 59, 121-2
covering strategies, 64-6, 84-5
credibility, 171-3

deadlines, 124-5
detectability, 105
deterrence, 14, 115-18, 144-7, 185-6, 198
  social-manufactured, 149-54, 160
discharges, nature of, 76-7, 81
discretion, xiv

efficacy, in standard setting, 24-6
efficiency, working notion of, 106, 135, 167, 198-202
enforcement behaviour, xii, xiv, 3-15, 16-17, 129ff
  as continuing relationship, 129ff
  as a game, 118-22, 144, 158-9, 172
enforcement career, 107, 164-6
equity,
  in prosecution decision-making, 181
  in standard setting, 24-6

farmers, 115, 136, 217, n.4
forbearance, 122-6, 167
formal sample *see* statutory samples

impact of pollution, 74-8
informants, cultivation of, 98-9
injunction, 131, 222 n.1
inspecting for pollution, 93-5

justice, conceptions of, 202-7

Kepone pollution, 210 n.26

law, images of, 187-90

moral categorization, 107, 109-18, 120-1, 132-3, 161-74, 197

negotiating tactics, 129ff
Notice of Intention to Commence Proceedings, 130, 182
noticeability of pollution, 75-8

'one-off' pollution, *see* pollution incidents
organizing data in prosecution cases, 183-7
organizations as objects of enforcement, 144-7

'persistent failures to comply', *see* persistent pollutions
persistent pollutions, 105-7, 129-30, 158-9, 162, 189, 206
polluters, images of, 110-18, 161-74
  'bolshie, 119-22
  'careless', 112
  'co-operative', 113-14, 164, 166-7, 171
  'malicious', 113
  'socially responsible', 110-12
  'troublemakers', 116-17
  'unco-operative', 114-15, 219 n. 14
  'unfortunate', 112
pollution
  as administrative creation, 23ff
  consequences of, xi, 12, 77-8

practical definition of, 72ff
as deviant occupation, 217 n.19
discovery of, 80-5
formal sanctions for, 11-12, 19, 151, 203
identification of, 85-90
legal definition of, 212 n.2
legislation, xii, 18-22
liability for, 13-14, 161-2, 169, 190-1, 203-6, 223 n.6
as organizational deviance, 14-15, 17
pollution control agencies
area offices, 50-3
area supervisors, 51-3
indices of impact, 195, 219, n.5
interested constituencies, 9-10, 193-5
vulnerability to public criticism of, 97-8, 193-8, 205-7
pollution control officers, 39ff
agency control of, 61-4, 70-1
as gatekeepers, 57-9
indices of output of, 63-4, 66-70
qualities required, 40-4, 50-3, 69-70, 82-3
role of, in prosecution, 183-7
pollution control work, 44-9
as autonomous activity, 57-61, 68-9
in contrast with policing, 59-60
knowledge required, 46
learning the job, 49-53, 215 n.20
nature of, 214, n.10, n.11
as policing, 44-6
pollution incidents, 65, 96, 105-7, 157, 189, 196, 206
pollution parameters, 29-30, 87
pollutions, images of, 112-13
'accidental', 112-13, 162, 164-5, 168-70
'blameless', 162, 168-9
'careless', 112-13, 163, 168
'deliberate', 162-3, 168, 170
'inevitable', 162, 197
'negligent', 162-3, 168-70
'serious', 130, 132
'technical', 197
proactive enforcement, 57, 73, 80-4, 90-6
prosecution, 130-1, 150-1, 160-1, 165-6, 170, 173, 177f, 191ff
levels of, 193-9
as organizational process, 179-87, 199
rates, explanations of, 191-3
statistics of, 177-8
as symbolic act, 194-5, 201-2
public stigma, 117-18
publicity, 84-5, 116-18, 180, 194-5, 205-6

reactive enforcement, 57, 90-1, 95-9
regional water authorities, 18-22
research methods, 225ff
access to agencies, 226
organization of fieldwork, 227-8
participant observation, 226, 228-9, 231
selection of agencies, 225, 227-8
selection of cases, 226-7
structured conversation, 229-30
validity of, 230-2
research perspective, 15-17
'Royal Commission' standards, 29-30, 79-80

sample analysis, 79-80
error in, 89-90
sample results, 88-90
sampling, 73, 85-93
sanctioning systems or strategies, 3-8, 66-7, 173-4, 182, 195
setting of pollution, 74-8
standard setting, 23ff
appeal to Secretary of State, 27-8
dilemmas in, 24-6
procedure followed, 28-32
standards, 23ff
as criteria of pollution, 78-80
enforceability of, 32-5
objectives of, 23-4
statutory samples, 63-5, 130, 155-61, 166, 174, 179, 181, 184, 213, n.14, 216, n.9
guidelines for, 156, 215, n.22
procedure followed, 155
suspiciousness, signs of, 83-4

threats, 137-40, 143, 158-61
'turning a blind eye', 73-4

Water Act, 1973
consequences of, 41, 53-6, 60-1